Introduction to AutoCAD 2011

Introduction to AutoCAD 2011

2D and 3D Design

Alf Yarwood

AMSTERDAM • BOSTON • HEIDELBERG • LONDON • NEW YORK • OXFORD
PARIS • SAN DIEGO • SAN FRANCISCO • SINGAPORE • SYDNEY • TOKYO

Newnes is an imprint of Elsevier

ELSEVIER

Newnes

Newnes is an imprint of Elsevier
The Boulevard, Langford Lane, Kidlington, Oxford OX5 1GB, UK
30 Corporate Drive, Suite 400, Burlington, MA 01803, USA

First edition 2010

British Library Cataloguing-in-Publication Data
A catalogue record for this book is available from the British Library

Library of Congress Cataloging-in-Publication Data
A catalog record for this book is available from the Library of Congress

ISBN: 978-0-08-096575-8

For information on all Newnes publications
visit our website at www.books.elsevier.com

Typeset by MPS Limited, a Macmillan Company, Chennai, India
www.macmillansolutions.com

Printed and bound in China

10 11 12 13 14 15   10 9 8 7 6 5 4 3 2 1

Contents

Preface

The purpose of writing this book is to produce a text suitable for students in Further and/or Higher Education who are required to learn how to use the computer-aided design (CAD) software package AutoCAD® 2011. Students taking examinations based on CAD will find the contents of the book of great assistance. The book is also suitable for those in industry wishing to learn how to construct technical drawings with the aid of AutoCAD 2011 and those who, having used previous releases of AutoCAD, wish to update their skills to AutoCAD 2011.

The chapters in Part 1 – 2D Design, dealing with two-dimensional (2D) drawing, will also be suitable for those wishing to learn how to use AutoCAD LT 2011, the 2D version of this latest release of AutoCAD.

Many readers using previous releases of AutoCAD will find the book's contents largely suitable for use with those versions, although AutoCAD 2011 has many enhancements over previous releases (some of which are mentioned in Chapter 21).

The contents of the book are basically a graded course of work, consisting of chapters giving explanations and examples of methods of constructions, followed by exercises which allow the reader to practise what has been learned in each chapter. The first 11 chapters are concerned with constructing technical drawing in 2D. These are followed by chapters detailing the construction of 3D solid drawings and rendering them. The final two chapters describe the Internet tools of AutoCAD 2011 and the place of AutoCAD in the design process. The book finishes with two appendices – a list of tools with their abbreviations and a list of some of the set variables upon which AutoCAD 2011 is based.

AutoCAD 2011 is very complex CAD software package. A book of this size cannot possibly cover the complexities of all the methods for constructing 2D and 3D drawings available when working with AutoCAD 2011. However, it is hoped that by the time the reader has worked through the contents of the book, he/she will be sufficiently skilled with methods of producing drawing with the software to be able to go on to more advanced constructions with its use and will have gained an interest in the more advanced possibilities available when using AutoCAD.

Alf Yarwood

Salisbury 2010 | **xiii**

Registered Trademarks

Autodesk® and AutoCAD® are registered in the US Patent and Trademark Office by Autodesk Inc.

Windows® is a registered trademark of the Microsoft Corporation.

Alf Yarwood is an Autodesk authorised author and a member of the Autodesk Advanced Developer Network.

2D Design

Introducing AutoCAD 2011

AIM OF THIS CHAPTER

The aim of this chapter is designed to introduce features of the AutoCAD 2011 window and methods of operating AutoCAD 2011.

CHAPTER 1

Opening AutoCAD 2011

Launch acad.exe

Fig. 1.1 The AutoCAD 2011 shortcut on the Windows desktop

AutoCAD 2011 is designed to work in a Windows operating system. In general, to open AutoCAD 2011, *double-click* on the **AutoCAD 2011** shortcut in the Windows desktop (Fig. 1.1). Depending on how details in **Profiles/Initial Setup…** in the **Options** dialog (Fig. 1.16, page 13), the **Welcome** dialog (Fig. 1.2) may appear. This dialog allows videos showing methods of working AutoCAD 2011, to be selected from a list of icons.

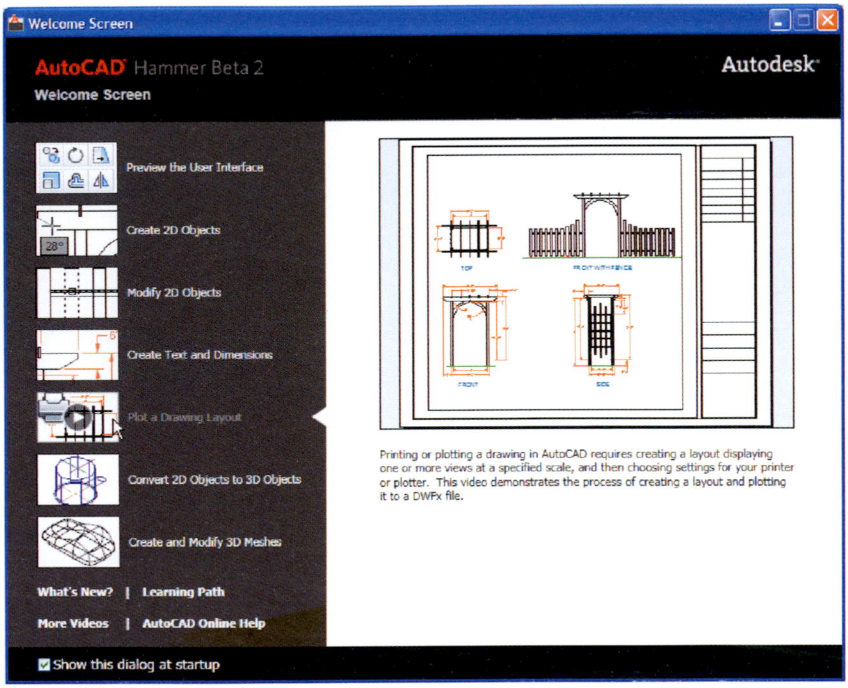

Fig. 1.2 Page 1 of the **Initial Settings** dialog

When working in education or in industry, computers may be configured to allow other methods of opening AutoCAD, such as a list appearing on the computer in use when the computer is switched on, from which the operator can select the program he/she wishes to use.

When AutoCAD 2011 is opened a window appears, which will depend upon whether a **3D Basics**, a **3D Modeling**, a **Classic AutoCAD** or a **2D Drafting & Annotation** workspace has been set as **QNEW** in the **Options** dialog. In this example the **2D Drafting & Annotation** workspace is shown and includes the **Ribbon** with **Tool panels** (Fig. 1.3). This **2D Drafting & Annotation** workspace shows the following details:

Ribbon: Which includes tabs, each of which when *clicked* will bring a set of panels containing tool icons. Further tool panels can be seen by

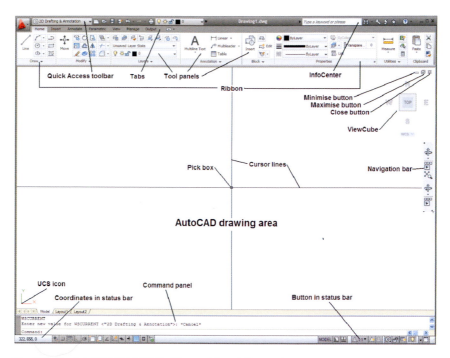

Fig. 1.3 The AutoCAD 2011 **2D Drafting and Annotation** workspace

clicking the appropriate tab. The panels in the ribbon can be changed to any desired panels as required using the **Customer User Interface** dialog if desired.

Menu Browser icon: A *left-click* on the arrow to the right of the **A** symbol at the top left-hand corner of the AutoCAD 2011 window causes the **Menu Browser** menu to appear (Fig. 1.4).

Workspace Switching menu: Appears with a *click* on the **Workspace Switching** button in the status bar (Fig. 1.5).

Command palette: Can be *dragged* from its position at the bottom of the AutoCAD window into the AutoCAD drawing area, when it can be seen to be a palette (Fig. 1.6). As with all palettes, an **Auto-hide** icon and a right-click menu is included.

Tool panels: Each shows tools appropriate to the panel. Taking the **Home/Draw** panel as an example, Fig. 1.7 shows that placing the mouse cursor on one of the tool icons in a panel brings a tooltip on screen showing details of how the tool can be used. Two types of tooltip will be seen. In the majority of future illustrations of tooltips, the smaller version will be shown. Other tools have popup menus appearing with a *click*. In the example given in Fig. 1.8, a *click* on the **Circle** tool icon will show a tooltip. A *click* on the arrow to the right of the tool icon brings a popup menu showing the construction method options available for the tool.

CHAPTER 1

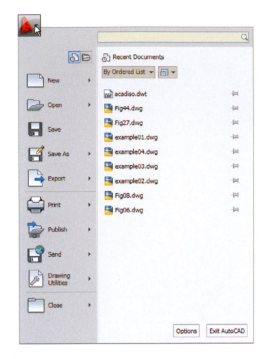

Fig. 1.4 The **Menu Browser**

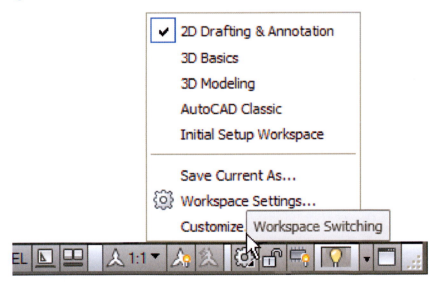

Fig. 1.5 The **Workspace Switching** popup menu

Fig. 1.6 The command palette when *dragged* from its position at the bottom of the AutoCAD window

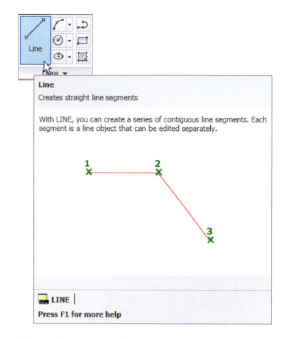

Fig. 1.7 The descriptive tooltip appearing with a *click* on the **Line** tool icon

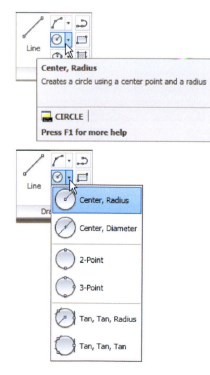

Fig. 1.8 The tooltip for the **Circle** tool and its popup menu

Quick Access toolbar: The toolbar at the top right of the AutoCAD window holds several icons, one of which is the **Open** tool icon. A *click* on the icon opens the **Select File** dialog (Fig. 1.9).

Navigation bar: contains several tools which may be of value.

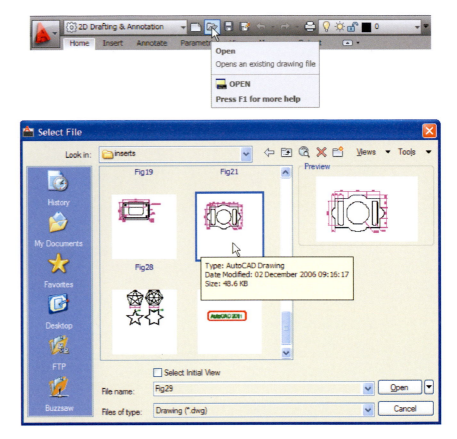

Fig. 1.9 The **open** icon in the **Quick Access** toolbar brings the **Select File** dialog on screen

The mouse as a digitiser

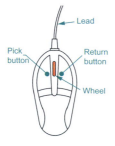

Fig. 1.10 The two-button mouse

Many operators working in AutoCAD will use a two-button mouse as a digitiser. There are other digitisers which may be used – pucks with tablets, a three-button mouse, etc. Fig. 1.10 shows a mouse which has two buttons and a wheel.

To operate this mouse pressing the **Pick button** is a *left-click*. Pressing the **Return button** is a *right-click* which usually, but not always, has the same result as pressing the **Enter** key of the keyboard.

When the **Wheel** is pressed drawings in the AutoCAD screen can be panned by moving the mouse. Moving the wheel forwards enlarges (zooms in) the drawing on screen. Move the wheel backwards and a drawing reduces in size.

The pick box at the intersection of the cursor hairs moves with the cursor hairs in response to movements of the mouse. The AutoCAD window as shown in Fig. 1.3 shows cursor hairs which stretch across the drawing in both horizontal and vertical directions. Some operators prefer cursor hairs to be shorter. The length of the cursor hairs can be adjusted in the **Display** sub-menu of the **Options** dialog (page 13).

Palettes

A palette has already been shown – the **Command** palette. Two palettes which may be frequently used are the **DesignCenter** palette and the **Properties** palette. These can be called to screen from icons in the **View/Palettes** panel.

DesignCenter palette: Fig. 1.11 shows the **DesignCenter** palette with the **Block** drawings of building symbols from which the block **Third type of chair** block has been selected.

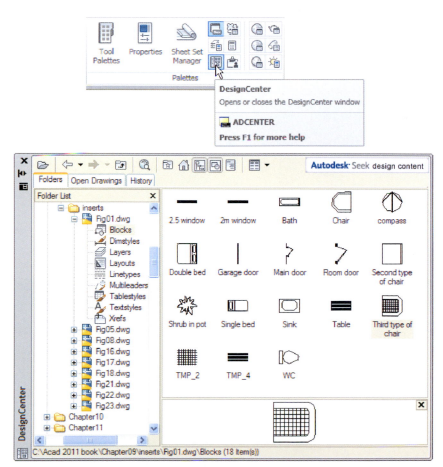

Fig. 1.11 A left-click on the **View/DesignCenter** icon brings the **DesignCenter** palette to screen

Properties palette: Fig. 1.12 shows the **Properties** palette, in which the general features of a selected line are shown. The line can be changed by *entering* new figures in parts of the palette.

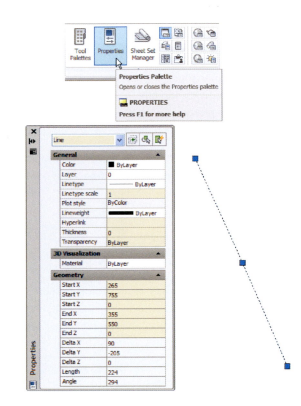

Fig. 1.12 The **Properties** palette

Tool palettes

Click on **Tool Palettes** in the **View/Palettes** panel and the **Tool Palettes – All Palettes** palette appears (Fig. 1.13).

Click in the title bar of the palette and a popup menu appears. *Click* on a name in the menu and the selected palette appears. The palettes can be reduced in size by *dragging* at corners or edges, or hidden by *clicking* on the **Auto-hide** icon, or moved by *dragging* on the **Move** icon. The palette can also be *docked* against either side of the AutoCAD window.

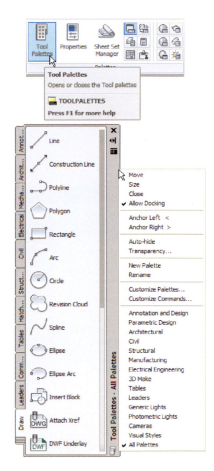

Fig. 1.13 The **Tool Palettes – All Palettes** palette

Notes

Throughout this book tools will often be shown as selected from the panels. It will be seen in Chapter 3 that tools can be 'called' in a variety of ways, but tools will frequently be shown selected from tool panels although other methods will also be shown on occasion.

Dialogs

Dialogs are an important feature of AutoCAD 2011. Settings can be made in many of the dialogs, files can be saved and opened, and changes can be made to variables.

Examples of dialogs are shown in Figs 1.15 and 1.16. The first example is taken from the **Select File** dialog (Fig. 1.15), opened with a *click* on **Open...** in the **Quick Access** toolbar (Fig. 1.14). The second example

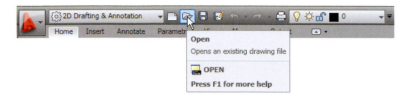

Fig. 1.14 Opening the **Select File** dialog from the **Open** icon in the **Quick Access** toolbar

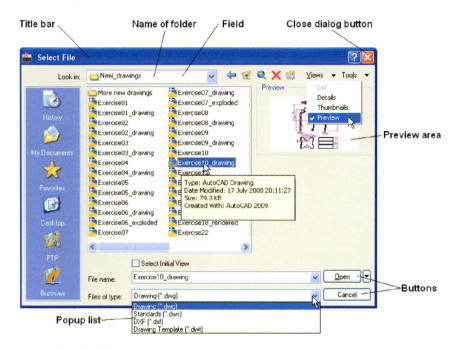

Fig. 1.15 The **Select File** dialog

shows part of the **Options** dialog (Fig. 1.16) in which many settings can be made to allow operators the choice of their methods of constructing drawings. The **Options** dialog can be opened with a *click* on **Options…** in the *right-click* dialog opened in the command palette.

Note the following parts in the dialog, many of which are common to other AutoCAD dialogs:

Title bar: Showing the name of the dialog.
Close dialog button: Common to other dialogs.
Popup list: A *left-click* on the arrow to the right of the field brings down a popup list listing selections available in the dialog.
Buttons: A *click* on the **Open** button brings the selected drawing on screen. A *click* on the **Cancel** button closes the dialog.
Preview area: Available in some dialogs – shows a miniature of the selected drawing or other feature, partly shown in Fig. 1.15.

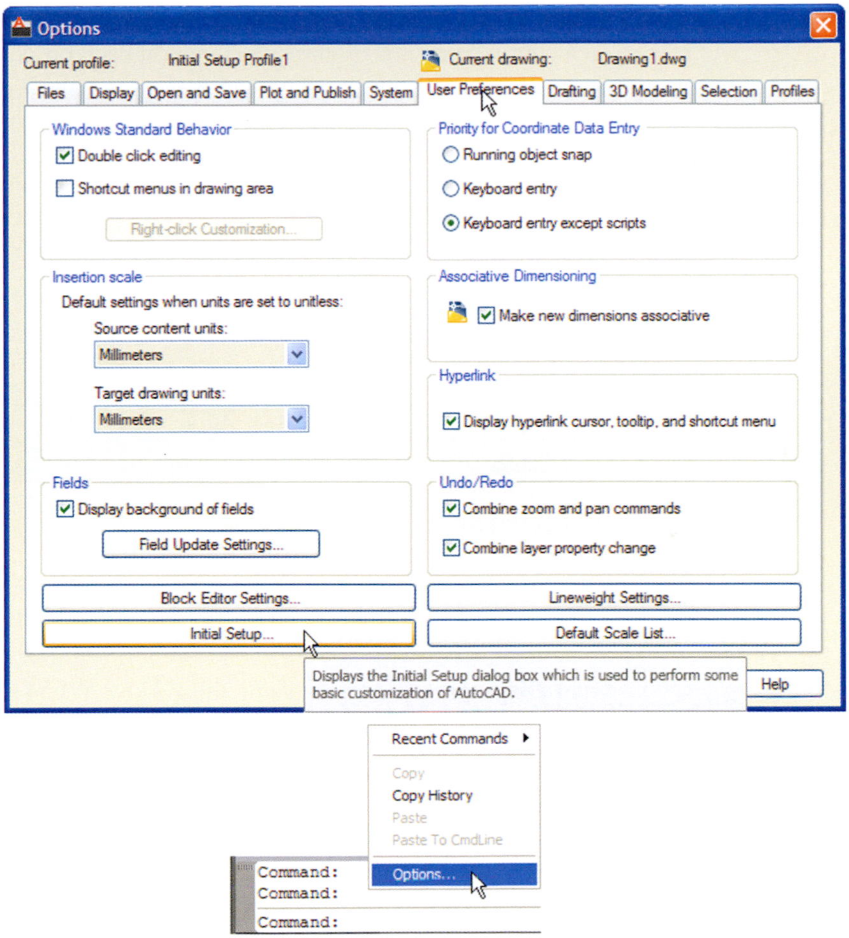

Fig. 1.16 Part of the **Options** dialog

Note the following in the **Options** dialog (Fig. 1.16):

Tabs: A *click* on any of the tabs in the dialog brings a sub-dialog on screen.

Check boxes: A tick appearing in a check box indicates the function described against the box is on. No tick and the function is off. A *click* in a check box toggles between the feature being off or on.

Radio buttons: A black dot in a radio button indicates the feature described is on. No dot and the feature is off.

Slider: A slider pointer can be *dragged* to change sizes of the feature controlled by the slider.

Buttons at the left-hand end of the status bar

A number of buttons at the left-hand end of the status bar can be used for toggling (turning on/off) various functions when operating within AutoCAD

2011 (Fig. 1.17). A *click* on a button turns that function on, if it is off; a *click* on a button when it is off turns the function back on. Similar results can be obtained by using function keys of the computer keyboard (keys **F1** to **F10**).

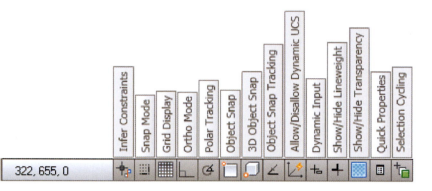

Fig. 1.17 The buttons at the left-hand end of the status bar

Snap Mode: Also toggled using the **F9** key. When snap on, the cursor under mouse control can only be moved in jumps from one snap point to another.

Grid Display: Also toggled using the **F7** key. When set on, a series of grid points appears in the drawing area.

Ortho Mode: Also toggled using the **F8** key. When set on, lines, etc. can only be drawn vertically or horizontally.

Polar Tracking: Also toggled using the **F10** key. When set on, a small tip appears showing the direction and length of lines, etc. in degrees and units.

Object Snap: Also toggled using the **F3** key. When set on, an osnap icon appears at the cursor pick box.

Object Snap Tracking: Also toggled by the **F11** key. When set on, lines, etc. can be drawn at exact coordinate points and precise angles.

Allow/Disallow Dynamic UCS: Also toggled by the **F6** key. Used when constructing 3D solid models.

Dynamic Input: Also toggled by **F12**. When set on, the **x,y** coordinates and prompts show when the cursor hairs are moved.

Show/Hide Lineweight: When set on, lineweights show on screen. When set off, lineweights only show in plotted/printed drawings.

Quick Properties: A *right-click* brings up a popup menu, from which a *click* on **Settings…** causes the **Drafting Settings** dialog to appear.

> **Note**
>
> When constructing drawings in AutoCAD 2011 it is advisable to toggle between **Snap**, **Ortho**, **Osnap** and the other functions in order to make constructing easier.

Buttons at the right-hand end of the status bar

Another set of buttons at the right-hand end of the status bar are shown in
Fig. 1.18. The uses of some of these will become apparent when reading future
pages of this book. A *click* on the downward-facing arrow near the right-hand
end of this set of buttons brings up the **Application Status Bar Menu**
(Fig. 1.19) from which the buttons in the status bar can be set on and/or off.

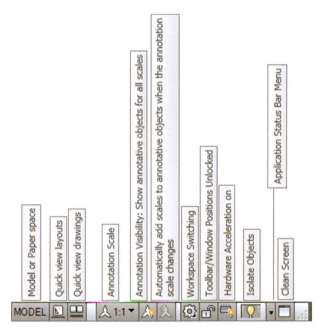

Fig. 1.18 The buttons at the right-hand end of the status bar

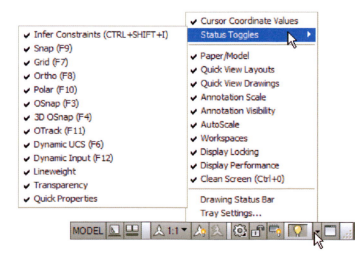

Fig. 1.19 The **Application Status Bar** menu

The AutoCAD coordinate system

In the AutoCAD 2D coordinate system, units are measured horizontally in terms of X and vertically in terms of Y. A 2D point in the AutoCAD drawing area can be determined in terms of X,Y (in this book referred to as *x,y*). *x,y* = 0,0 is the **origin** of the system. The coordinate point *x,y* = 100,50 is 100 units to the right of the origin and 50 units above the origin. The point *x,y* = −100, −50 is 100 units to the left of the origin and 50 points below the origin. Fig. 1.20 shows some 2D coordinate points in the AutoCAD window.

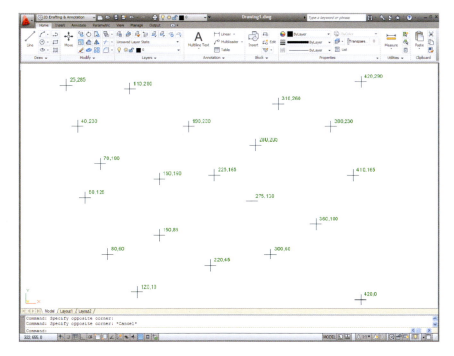

Fig. 1.20 The 2D coordinate points in the AutoCAD coordinate system

3D coordinates include a third coordinate (Z), in which positive Z units are towards the operator as if coming out of the monitor screen and negative Z units going away from the operator as if towards the interior of the screen. 3D coordinates are stated in terms of *x,y,z*. *x,y,z* = 100,50,50 is 100 units to the right of the origin, 50 units above the origin and 50 units towards the operator. A 3D model drawing as if resting on the surface of a monitor is shown in Fig. 1.21.

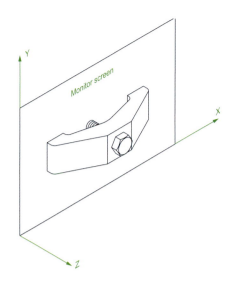

Fig. 1.21 A 3D model drawing showing the X, Y and Z coordinate directions

Drawing templates

Drawing templates are files with an extension **.dwt**. Templates are files
which have been saved with predetermined settings – such as **Grid** spacing
and **Snap** spacing. Templates can be opened from the **Select template**
dialog (Fig. 1.22) called by *clicking* the **New...** icon in the **Quick Access**

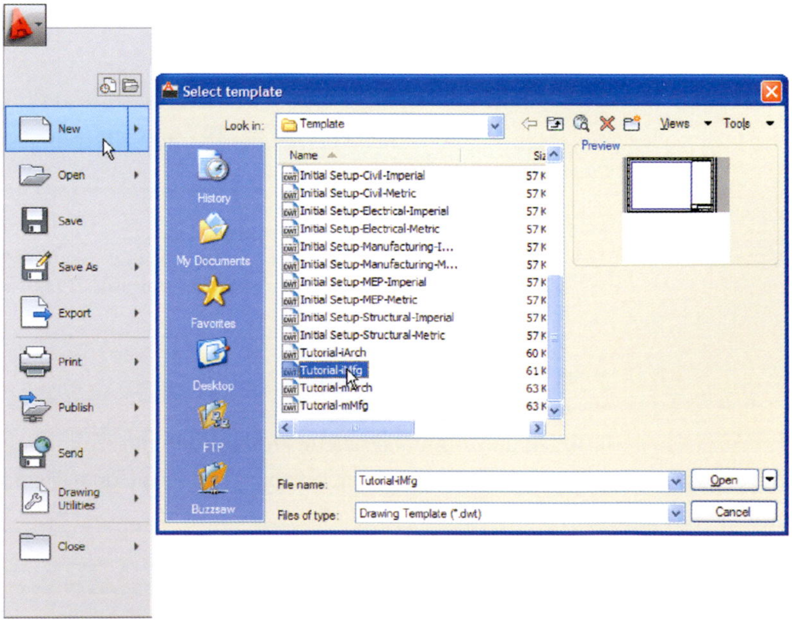

Fig. 1.22 A template selected from the **Select template** dialog

toolbar. An example of a template file being opened is shown in Fig. 1.22. In this example the template will be opened in Paper Space and is complete with a title block and borders.

When AutoCAD 2011 is used in European countries and opened, the **acadiso.dwt** template is the one most likely to appear on screen. In this part (Part 1 – 2D Design) of this book drawings will usually be constructed in an adaptation of the **acadiso.dwt** template. To adapt this template:

1. In the command palette *enter* (type) **grid** followed by a *right-click* (or pressing the **Enter** key). Then *enter* **10** in response to the prompt which appears, followed by a *right-click* (Fig. 1.23).

```
Command: grid
Specify grid spacing(X) or [ON/OFF/Snap/Major/aDaptive/Limits/
Follow/Aspect]<0>: 10
Command:
```

Fig. 1.23 Setting **Grids** to **10**

2. In the command palette *enter* **snap** followed by *right-click*. Then *enter* **5** followed by a *right-click* (Fig. 1.24).

```
Command: snap
Specify snap spacing or [ON/OFF/Aspect/Style/Type]<10>:5
Command:
```

Fig. 1.24 Setting **Snap** to **5**

3. In the command palette *enter* **limits**, followed by a *right-click*. *Right-click* again. Then *enter* **420, 297** and *right-click* (Fig. 1.25).

```
Command: limits
Reset Model space limits:
Specify lower left corner or [ON/OFF] <0,0>:
Specify upper right corner <12,9>:420,297
Command:
```

Fig. 1.25 Setting **Limits** to **420, 297**

4. In the command palette *enter* **zoom** and *right-click*. Then in response to the line of prompts which appears *enter* **a** (for All) and *right-click* (Fig. 1.26).

```
Command: zoom
Specify corner of window, enter a scale factor (nX or nXP), or
[All/Center/Dynamic/Extents/Previous/Scale/Window/Object] <real time>: a
Regenerating model.
Command:
```

Fig. 1.26 Zooming to All

5. In the command palette *enter* **units** and *right-click*. The **Drawing Units** dialog appears (Fig. 1.27). In the **Precision** popup list of the **Length** area of the dialog, *click* on **0** and then *click* the **OK** button. Note the change in the coordinate units showing in the status bar.

Fig. 1.27 Setting **Units** to **0**

6. *Click* the **Save** icon in the **Quick Access** toolbar (Fig. 1.28). The **Save Drawing As** dialog appears. In the **Files of type** popup list select **AutoCAD Drawing Template (*.dwt)**. The templates already in AutoCAD are displayed in the dialog. *Click* on **acadiso.dwt**, followed by another *click* on the **Save** button.

Fig. 1.28 Click **Save**

Notes

1. Now when AutoCAD is opened the template saved as **acadiso.dwt** automatically loads with **Grid** set to **10**, **Snap** set to **5**, **Limits** set to **420,297** (size of an A3 sheet in millimetres) and with the drawing area zoomed to these limits, with **Units** set to **0**.

2. However, if there are multiple users by the computer, it is advisable to save your template to another file name, e.g. **my_template.dwt**.

3. Other features will be added to the template in future chapters.

Methods of showing entries in the command palette

Throughout the book, a tool is "called" usually by a *click* on a tool icon in a panel – in this example *entering* **zoom** at the command line and the following appears in the command palette:

```
Command: enter zoom right-click
Specify corner of window, enter a scale factor
   (nX or nXP), or [All/Center/Dynamic/Extents/
   Previous/Scale/Window/Object] <real time>: pick
   a point on screen
Specify opposite corner: pick another point to
   form a window
Command:
```

> **Note**
>
> In later examples this may be shortened to:
>
> ```
> Command: zoom
> [prompts]: following by picking points
> Command:
> ```

> **Notes**
>
> 1. In the above *enter* means type the given letter, word or words at the **Command:** prompt.
>
> 2. *Right-click* means press the **Return** (right) button of the mouse or press the **Return** key of the keyboard.

Tools and tool icons

In AutoCAD 2011, tools are shown as names and icons in panels or in drop-down menus. When the cursor is placed over a tool icon a description shows with the name of the tool as shown and an explanation in diagram form as in the example given in Fig. 1.7 (page 5).

If a small outward-facing arrow is included at the right-hand side of a tool icon, when the cursor is placed over the icon and the *pick* button of the mouse depressed and held, a flyout appears which includes other features. An example is given in Fig. 1.8 (page 5).

Another AutoCAD workspace

Other workspaces can be selected as the operator wishes. One in particular which may appeal to some operators is to *click* **AutoCAD Classic** in the **2D Drafting & Annotation** popup menu (Fig. 1.29).

Fig. 1.30 shows the **AutoCAD Classic** workspace screen.

Fig. 1.29 Selecting **Classic Workspace** from the popup menu

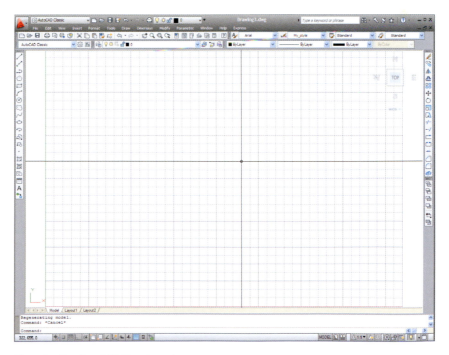

Fig. 1.30 The **AutoCAD Classic** workspace

In the **AutoCAD Classic** workspace, tools icons are held in toolbars, which are *docked* against the sides and top of the workspace. The tool icons in the **Draw** toolbar (*docked* left-hand side) are shown in Fig. 1.31. Note the grid lines, spaced at **10** coordinate units in both **X** and **Y** directions.

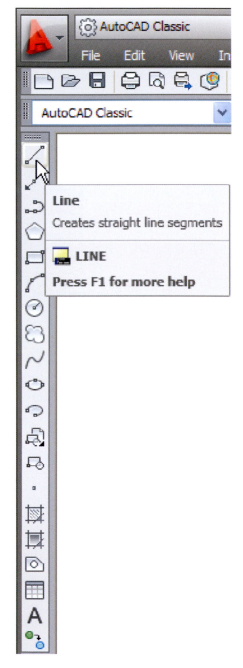

Fig. 1.31 The tool icons in the Draw toolbar

The Ribbon

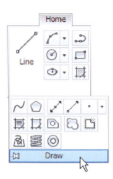

Fig. 1.32 The **Home/ Draw** panel and its flyout

In the **2D Drafting & Annotation** workspace, the **Ribbon** contains groups of panels placed at the top of the AutoCAD 2011 window. In Fig. 1.3 on page 3, there are eight panels – **Draw**, **Modify**, **Layers**, **Annotation**, **Block**, **Properties**, **Utilities** and **Clipboard**. Other groups of palettes can be called from the **tabs** at the top of the **Ribbon**.

If a small arrow is showing below the panel name, a *left-click* on the arrow brings down a flyout showing additional tool icons in the panel. As an example Fig. 1.32 shows the flyout from the **Home/Draw** panel.

At the right-hand end of the panel titles (the **tabs**) are two downward pointing arrows. A *left-click* on the right of these two arrows brings down a menu. A *right-click* on the same arrow brings down another menu (Fig. 1.33). Options from these two menus show that the ribbon can

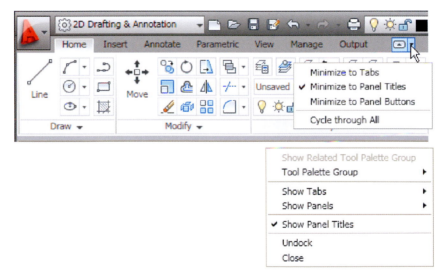

Fig. 1.33 The two menus from the right-hand arrow

appear in the AutoCAD window in a variety of ways. It is worth while experimenting with the settings of the ribbon – each operator will find the best for him/herself. The left-hand arrow also varies the ribbon.

Repeated *left-clicks* on this arrow cause the **Ribbon** panels to:

1. Minimize to tabs
2. Minimize to panel titles
3. Minimize to panel button
4. The full ribbon.

CHAPTER 1

Continuing *clicks* cause the changes to revert to the previous change.

Fig. 1.34 shows the **Minimize** settings. Any one of these settings leaves more space in the AutoCAD drawing window in which to construct drawings. The various settings of the ribbon allow the user discretion as to how to use the ribbon. When minimized to panel titles or to panel buttons passing the cursor over the titles or buttons causes the panels to reappear and allow selection of tools. Also try **Undock** from the *right-click* menu.

Minimize to tabs

Minimize to panel titles

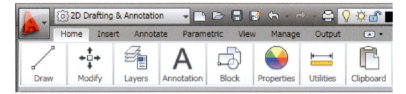

Minimize to panel buttons

Fig. 1.34 The **Ribbon** minimize settings

The Quick View Drawings button

One of the buttons at the right-hand end of the status bar is the **Quick View Drawings** button. A *click* on this button brings miniatures of recent drawings on screen (Fig. 1.35). This can be of value when wishing to check back features of recent drawings in relation to the current drawing on screen.

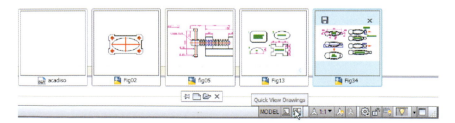

Fig. 1.35 The result of a click on the **Quick View Drawings** button

Customisation of the User Interface

The AutoCAD 2011 workspace can be arranged in any format the operator wishes by making settings in the **Customize User Interface** dialog (Fig. 1.36) brought to screen from the *right-click* menu from the button in the **Quick Access** toolbar. The dialog can be opened using other methods such as *entering* **cui** at the command line, but using this *right-click* menu is possibly the quickest method. The dialog is only shown here to alert the reader to the fact that he/she can customise the workspace being used to suit their own methods of working. Page space in this book does not allow further explanation.

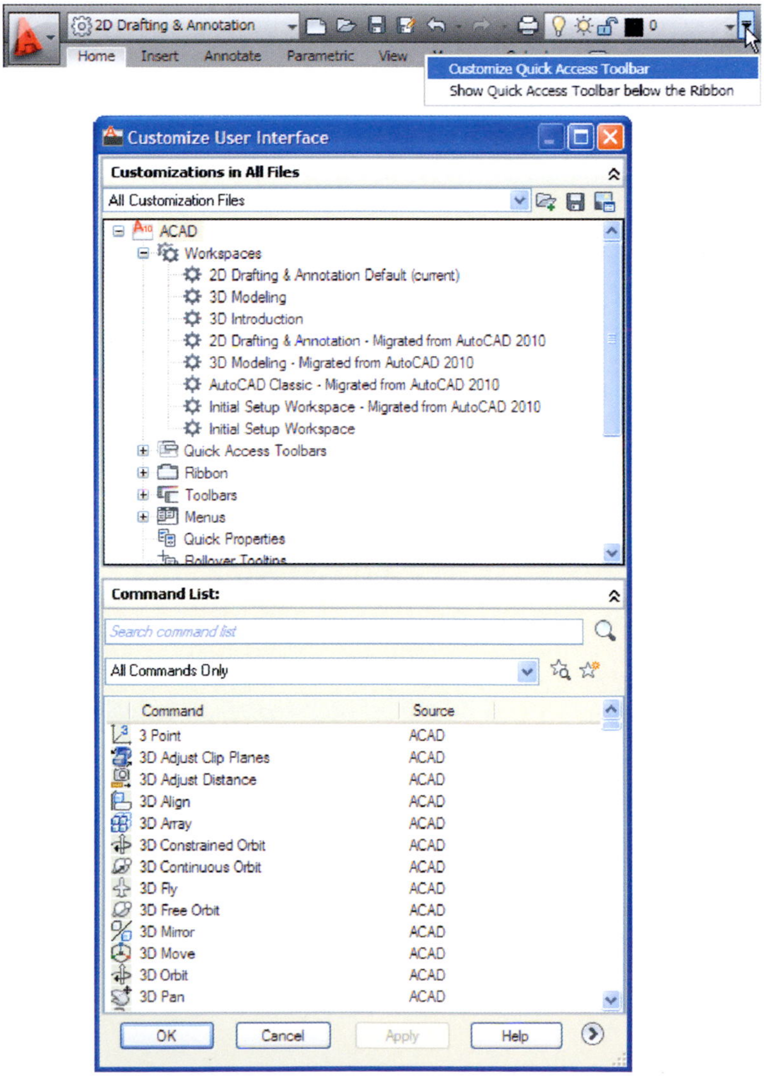

Fig. 1.36 The **Customize User Interface** dialog

REVISION NOTES

1. A *double-click* on the AutoCAD 2011 shortcut in the Windows desktop opens the AutoCAD window.
2. There are FOUR main workspaces in which drawings can be constructed – the 2D Drafting & Annotation, AutoCAD Classic, 3D Basics, 3D Modeling. Part 1, 2D Design, of this book deals with 2D drawings and these will be constructed mainly in the 2D Drafting & Annotation workspace. In Part 2, 3D Design, 3D model drawings will be mainly constructed in the 3D Modeling workspace.
3. All constructions in this book involve the use of a mouse as the digitiser. When a mouse is the digitiser:
   A *left-click* means pressing the left-hand button (the Pick) button.
   A *right-click* means pressing the right-hand button (the Return) button.
   A *double-click* means pressing the left-hand button twice in quick succession.
   Dragging means moving the mouse until the cursor is over an item on screen, holding the left-hand button down and moving the mouse. The item moves in sympathy to the mouse movement.
   To *pick* has a similar meaning to a *left-click*.
4. Palettes are a particular feature of AutoCAD 2011. The Command palette and the DesignCenter palette will be in frequent use.
5. Tools are shown as icons in the tool panels.
6. When a tool is picked, a tooltip describing the tool appears describing the action of the tool. Tools show a small tooltip, followed shortly afterwards by a larger one, but the larger one can be prevented from appearing by selecting an option in the Options dialog.
7. Dialogs allow opening and saving of files and the setting of parameters.
8. A number of *right-click* menus are used in AutoCAD 2011.
9. A number of buttons in the status bar can be used to toggle features such as snap and grid. Functions keys of the keyboard can be also used for toggling some of these functions.
10. The AutoCAD coordinate system determines the position in units of any 2D point in the drawing area (2D Drafting & Annotation) and any point in 3D space (3D Modeling).
11. Drawings are usually constructed in templates with predetermined settings. Some templates include borders and title blocks.

Note

Throughout this book when tools are to be selected from panels in the ribbon the tools will be shown in the form, e.g. **Home/Draw** – the name of the tab in the ribbon title bar, followed by the name of the panel from which the tool is to be selected.

Introducing drawing

AIMS OF THIS CHAPTER

The aims of this chapter are:

1. To introduce the construction of 2D drawing in the **2D Drafting & Annotation** workspace.
2. The drawing of outlines using the **Line**, **Circle** and **Polyline** tools from the **Home/Draw** panel.
3. Drawing to snap points.
4. Drawing to absolute coordinate points.
5. Drawing to relative coordinate points.
6. Drawing using the 'tracking' method.
7. The use of the **Erase**, **Undo** and **Redo** tools.

The 2D Drafting & Annotation workspace

Illustrations throughout this chapter will be shown as working in the **2D Drafting & Annotation** workspace. In this workspace the **Home/Draw** panel is at the left-hand end of the **Ribbon**, and **Draw** tools can be selected from the panel as indicated by a *click* on the **Line** tool (Fig. 2.1). In this chapter all examples will show tools as selected from the **Home/Draw** panel. However, methods of construction will be the same if the reader wishes to work by calling tools from the **Draw** drop-down menu. In order to bring drop-down menus on screen, first *click* the small arrow button on the right-hand end of the **Quick Access** toolbar, then *click* **Show Menu Bar** in the menu which appears. Menu titles appear above the **Ribbon**. *Click* **Draw** in this menu bar. From the drop-down menu which appears tools from the **Draw** list in the menu can be selected. Fig. 2.2 shows the **Line** tool being selected.

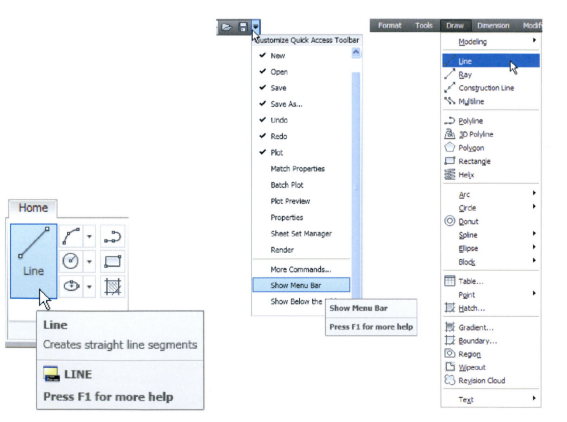

Fig. 2.1 The **Line** tool from the **Home/Draw** Panel with its tooltip

Fig. 2.2 Selecting the **Line** tool in the **2D Drafting & Annotation** workspace

CHAPTER 2

Drawing with the Line tool

First example – Line tool (Fig. 1.3)

1. Open AutoCAD. The drawing area will show the settings of the **acadiso.dwt** template – **Limits** set to **420,297**, **Grid** set to **10**, **Snap** set to **5** and **Units** set to **0**.

2. *Left-click* on the **Line** tool in the **Home/Draw** panel (Fig. 2.1), or *click* **Line** in the **Draw** drop-down menu (Fig. 2.2), or *enter* **line** or **l** at the command line.

> **Notes**
>
> a. The tooltip which appears when the tool icon is *clicked* in the **Draw** panel.
>
> b. The prompt **Command:_line Specify first point** which appears in the command window at the command line (Fig. 2.3). 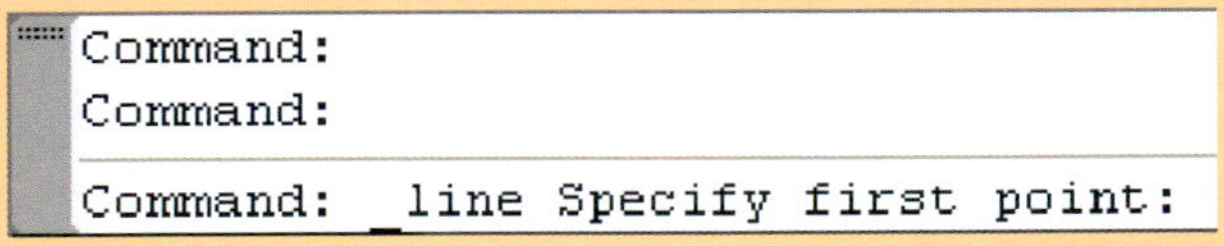
>
> ```
> Command:
> Command:
> Command:  _line Specify first point:
> ```
>
> **Fig. 2.3** The prompt appearing at the command line in the Command palette when **Line** is 'called'

3. Make sure **Snap** is on by either pressing the **F9** key or the **Snap Mode** button in the status bar. <**Snap on**> will show in the command palette.

4. Move the mouse around the drawing area. The cursors pick box will jump from point to point at 5 unit intervals. The position of the pick box will show as coordinate numbers in the status bar (left-hand end).

5. Move the mouse until the coordinate numbers show **60,240,0** and press the *pick* button of the mouse (*left-click*).

6. Move the mouse until the coordinate numbers show **260,240,0** and *left-click*.

7. Move the mouse until the coordinate numbers show **260,110,0** and *left-click*.

8. Move the mouse until the coordinate numbers show **60,110,0** and *left-click*.

9. Move the mouse until the coordinate numbers show **60,240,0** and *left-click*. Then press the **Return** button of the mouse (*right-click*).

The line rectangle Fig. 2.4 appears in the drawing area.

Fig. 2.4 First example – **Line** tool

Second example – **Line tool** (Fig. 2.6)

1. Clear the drawing from the screen with a *click* on the **Close** button of the AutoCAD drawing area. Make sure it is not the AutoCAD 2011 window button.
2. The warning window Fig. 2.5 appears in the centre of the screen. *Click* its **No** button.

Fig. 2.5 The **AutoCAD** warning window

3. *Left-click* **New…** button in the **File** drop-down menu and from the **Select template** dialog which appears *double-click* on **acadiso.dwt**.
4. *Left-click* on the **Line** tool icon and *enter* figures as follows at each prompt of the command line sequence:

```
Command:_line Specify first point: enter 80,235
  right-click
Specify next point or [Undo]: enter 275,235
right-click
Specify next point or [Undo]: enter 295,210
  right-click
```

```
Specify next point or [Close/Undo]:  enter 295,100
  right-click
Specify next point or [Close/Undo]:  enter 230,100
  right-click
Specify next point or [Close/Undo]:  enter 230,70
  right-click
Specify next point or [Close/Undo]:  enter 120,70
  right-click
Specify next point or [Close/Undo]:  enter 120,100
  right-click
Specify next point or [Close/Undo]:  enter 55,100
  right-click
Specify next point or [Close/Undo]:  enter 55,210
  right-click
Specify next point or [Close/Undo]:  enter c
  (Close) right-click
Command:
```

The result is as shown in Fig. 2.6.

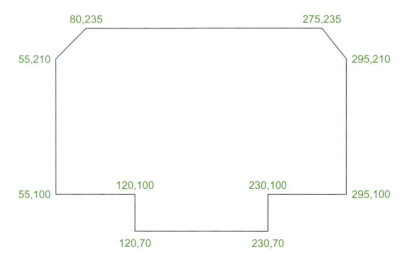

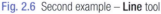

Fig. 2.6 Second example – **Line** tool

Third example – Line tool (Fig. 2.7)

1. Close the drawing and open a new **acadiso.dwt** window.
2. *Left-click* on the **Line** tool icon and *enter* figures as follows at each prompt of the command line sequence:

```
Command:_line Specify first point:  enter 60,210
  right-click
```

```
Specify next point or [Undo]: enter @50,0
  right-click
Specify next point or [Undo]: enter @0,20
  right-click
Specify next point or [Close/Undo]: enter @130,0
  right-click
Specify next point or [Close/Undo]: enter @0,-20
  right-click
```

Third example – Line tool (Fig. 2.7)

1. Close the drawing and open a new **acadiso.dwt** window.
2. *Left-click* on the **Line** tool icon and *enter* figures as follows at each prompt of the command line sequence:

```
Command:_line Specify first point: enter 60,210
  right-click
Specify next point or [Undo]: enter @50,0
  right-click
Specify next point or [Undo]: enter @0,20
  right-click
Specify next point or [Undo/Undo]: enter @130,0
  right-click
Specify next point or [Undo/Undo]: enter @0,-20
  right-click
Specify next point or [Undo/Undo]: enter @50,0
  right-click
Specify next point or [Close/Undo]: enter @0,-105
  right-click
Specify next point or [Close/Undo]: enter @-50,0
  right-click
Specify next point or [Close/Undo]: enter @0,-20
  right-click
Specify next point or [Close/Undo]: enter @-130,0
  right-click
Specify next point or [Close/Undo]: enter @0,20
  right-click
Specify next point or [Close/Undo]: enter @-50,0
  right-click
Specify next point or [Close/Undo]: enter c
  (Close) right-click
Command:
```

The result is as shown in Fig. 2.7.

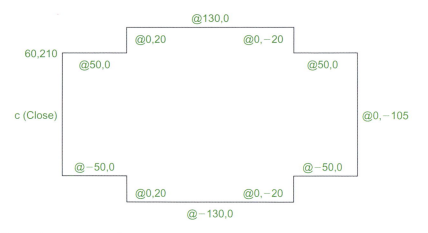

Fig. 2.7 Third example – **Line** tool

Notes

1. The figures typed at the keyboard determining the corners of the outlines in the above examples are two-dimensional (2D) **x,y** coordinate points. When working in 2D, coordinates are expressed in terms of two numbers separated by a comma.

2. Coordinate points can be shown in positive or negative numbers.

3. The method of constructing an outline as shown in the first two examples above is known as the **absolute coordinate entry** method, where the **x,y** coordinates of each corner of the outlines are *entered* at the command line as required.

4. The method of constructing an outline as in the third example is known as the **relative coordinate entry** method – coordinate points are *entered* relative to the previous entry. In relative coordinate entry, the @ symbol is *entered* before each set of coordinates with the following rules in mind:

    +ve x entry is to the right.

    −ve x entry is to the left.

    +ve y entry is upwards.

    −ve y entry is downwards.

5. The next example (the fourth) shows how lines at angles can be drawn taking advantage of the relative coordinate entry method. Angles in AutoCAD are measured in 360 degrees in a

counterclockwise (anticlockwise) direction (Fig. 2.8). The < symbol precedes the angle.

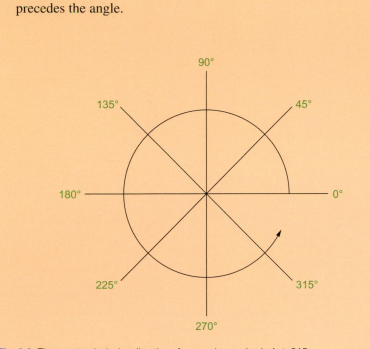

Fig. 2.8 The counterclockwise direction of measuring angles in AutoCAD

Fourth example – Line tool (Fig. 2.9)

1. Close the drawing and open a new **acadiso.dwt** window.
2. *Left-click* on the **Line** tool icon and *enter* figures as follows at each prompt of the command line sequence:

```
Command:_line Specify first point: 70,230
Specify next point: @220,0
Specify next point: @0,-70
Specify next point or [Undo]: @115<225
Specify next point or [Undo]: @-60,0
Specify next point or [Close/Undo]: @115<135
Specify next point or [Close/Undo]: @0,70
Specify next point or [Close/Undo]: c (Close)
Command:
```

The result is as shown in Fig. 2.9.

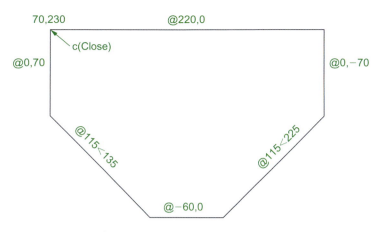

Fig. 2.9 Fourth example – **Line** tool

Fifth example – Line tool (Fig. 2.10)

Another method of constructing accurate drawings is by using a method
known as **tracking**. When **Line** is in use, as each **Specify next point:**
appears at the command line, a *rubber-banded* line appears from the last
point *entered*. *Drag* the rubber-band line in any direction and *enter* a
number at the keyboard, followed by a *right-click*. The line is drawn in the
dragged direction of a length in units equal to the *entered* number.

In this example because all lines are drawn in vertical or horizontal
directions, either press the **F8** key or *click* the **ORTHO** button in the status
bar which will only allow drawing horizontally or vertically.

1. Close the drawing and open a new **acadiso.dwt** window.
2. *Left-click* on the **Line** tool icon and *enter* figures as follows at each
   prompt of the command line sequence:

```
Command:_line Specify first point: enter 65,220
  right-click
Specify next point: drag to right enter 240
  right-click
Specify next point: drag down enter 145 right-click
Specify next point or [Undo]: drag left enter 65
  right-click
Specify next point or [Undo]: drag upwards enter 25
  right-click
Specify next point or [Close/Undo]: drag left
  enter 120 right-click
Specify next point or [Close/Undo]: drag upwards
  enter 25 right-click
```

CHAPTER 2

```
Specify next point or [Close/Undo]: drag left
   enter 55 right-click
Specify next point or [Close/Undo]: c (Close)
   right-click
Command:
```

The result is as shown in Fig. 2.10.

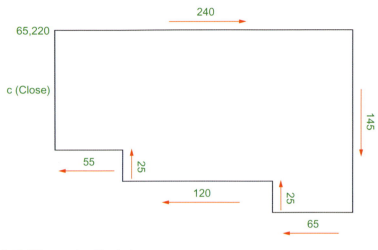

Fig. 2.10 Fifth example – **Line** tool

Drawing with the Circle tool

First example – Circle tool (Fig. 2.13)

1. Close the drawing just completed and open the **acadiso.dwt** template.
2. *Left-click* on the **Circle** tool icon in the **Home/Draw** panel (Fig. 2.11).

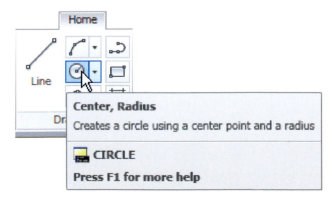

Fig. 2.11 The **Circle** tool from the **Home/Draw** panel

3. *Enter* a coordinate and a radius against the prompts appearing in the command window as shown in Fig. 2.12, followed by *right-clicks*. The circle (Fig. 2.13) appears on screen.

```
Command: _circle Specify center point for circle or [3P/2P/Ttr (tan tan
radius)]: 180,160
Specify radius of circle or [Diameter]: 55
Command:
```

Fig. 2.12 First example – **Circle**. The command line prompts when **Circle** is called

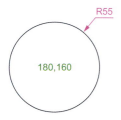

Fig. 2.13 First example – **Circle** tool

Second example – Circle tool (Fig. 2.15)

1. Close the drawing and open the **acadiso.dwt** screen.
2. *Left-click* on the **Circle** tool icon and construct two circles as shown in the drawing Fig. 2.14 in the positions and radii shown in Fig. 2.15.

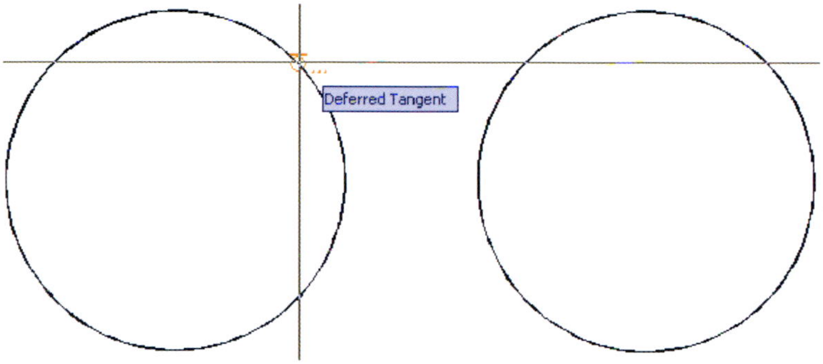

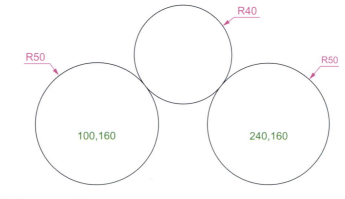

Deferred Tangent

Fig. 2.14 Second example – **Circle** tool – the two circles of radius 50

Fig. 2.15 Second example

3. *Click* the **Circle** tool again and against the first prompt *enter* **t** (the abbreviation for the prompt **tan tan radius**), followed by a *right-click*.

```
Command_circle Specify center point for circle or
   [3P/2P/Ttr (tan tan radius]: enter t right-click
Specify point on object for first tangent of
   circle: pick
Specify point on object for second tangent of
   circle: pick
Specify radius of circle (50): enter 40 right-
   click
Command:
```

The circle of radius **40** tangential to the two circles already drawn then appears (Fig. 2.15).

> **Notes**
>
> 1. When a point on either circle is picked a tip (**Deferred Tangent**) appears. This tip will only appear when the **Object Snap** button is set on with a *click* on its button in the status bar, or the **F3** key of the keyboard is pressed.
>
> 2. Circles can be drawn through 3 points or through 2 points *entered* at the command line in response to prompts brought to the command line by using **3P** and **2P** in answer to the circle command line prompts.

The Erase tool

If an error has been made when using any of the AutoCAD 2011 tools, the object or objects which have been incorrectly drawn can be deleted with the **Erase** tool. The **Erase** tool icon can be selected from the **Home/Modify** panel (Fig. 2.16) or by *entering* **e** at the command line.

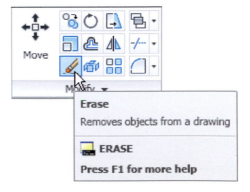

Fig. 2.16 The **Erase** tool icon from the **Home/Modify** panel

First example – Erase (Fig. 2.18)

1. With **Line** construct the outline Fig. 2.17.

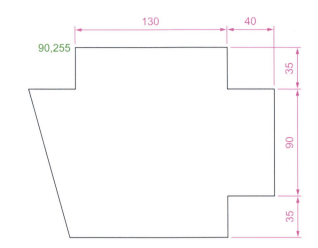

Fig. 2.17 First example – **Erase**. An incorrect outline

2. Assuming two lines of the outline have been incorrectly drawn, *left-click* the **Erase** tool icon. The command line shows:

```
Command:_erase
Select objects: pick one of the lines 1 found
Select objects: pick the other line 2 total
Select objects: right-click
Command:
```

And the two lines are deleted (right-hand drawing of Fig. 2.18).

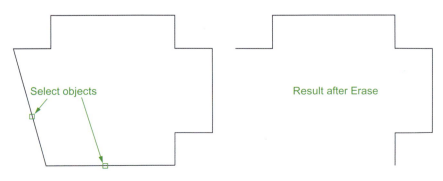

Select objects

Result after Erase

Fig. 2.18 First example – **Erase**

Second example – Erase (Fig. 2.19)

The two lines could also have been deleted by the following method:

1. *Left-click* the **Erase** tool icon. The command line shows:

```
Command:_erase
Select objects: enter c (Crossing)
Specify first corner: pick Specify opposite corner:
 pick 2 found
Select objects: right-click
Command:
```

And the two lines are deleted as in the right-hand drawing Fig. 2.19.

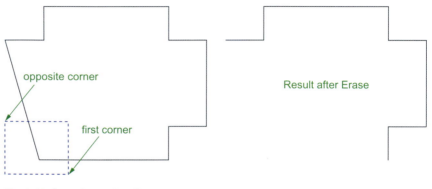

Fig. 2.19 Second example – **Erase**

Undo and Redo tools

Two other tools of value when errors have been made are the **Undo** and **Redo** tools. To undo any last action when constructing a drawing, either *left-click* the **Undo** tool in the **Quick Access** toolbar (Fig. 2.20) or *enter* **u** at the command line. No matter which method is adopted the error is deleted from the drawing.

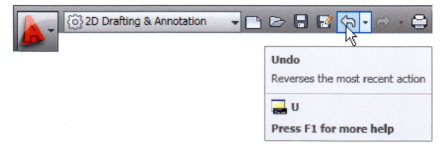

Fig. 2.20 The **Undo** tool in the **Quick Access** toolbar

Everything constructed during a session of drawing can be undone by repeated *clicking* on the **Undo** tool icon or by repeatedly *entering* **u**'s at the command line.

To bring back objects that have just been removed by the use of **Undo**'s, *left-click* the **Redo** tool icon in the **Quick Access** toolbar (Fig. 2.21) or *enter* **redo** at the command line.

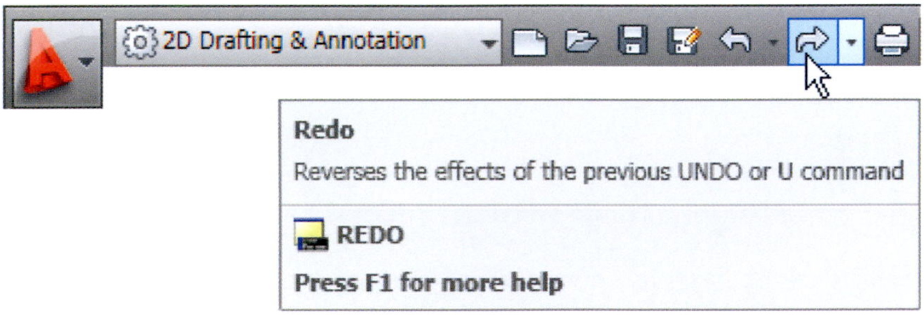

Fig. 2.21 The **Redo** tool icon in the **Quick Access** toolbar

Drawing with the Polyline tool

When drawing lines with the **Line** tool, each line drawn is an object. A rectangle drawn with the **Line** tool is four objects. A rectangle drawn with the **Polyline** tool is a single object. Lines of different thickness, arcs, arrows and circles can all be drawn using this tool. Constructions resulting from using the tool are known as **polylines** or **plines**. The tool can be called from the **Home/Draw** panel (Fig. 2.22) or by *entering* **pl** at the command line.

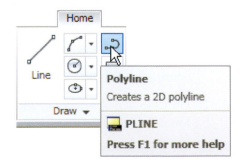

Fig. 2.22 The **Polyline** tool icon in the **Home/Draw** panel

First example – Polyline tool (Fig. 2.23)

In this example *enter* and *right-click* have not been included (Fig. 2.23).

Left-click the **Polyline** tool icon. The command line shows:

```
Command:_pline Specify start point: 30,250
Current line width is 0
Specify next point or [Arc/Halfwidth/Length/Undo/
  Width]: 230,250
Specify next point or [Arc/Close/Halfwidth/Length/
  Undo/Width]: 230,120
Specify next point or [Arc/Close/Halfwidth/Length/
  Undo/Width]: 30,120
Specify next point or [Arc/Close/Halfwidth/Length/
  Undo/Width]: c (Close)
Command:
```

Fig. 2.23 First example – **Polyline** tool

Notes

1. Note the prompts – **Arc** for constructing pline arcs, **Close** to close an outline, **Halfwidth** to halve the width of a wide pline, **Length** to *enter* the required length of a pline, **Undo** to undo the last pline constructed **Width** to change the width of the pline.

2. Only the capital letter(s) of a prompt needs to be *entered* in upper or lower case to make that prompt effective.

3. Other prompts will appear when the **Polyline** tool is in use as will be shown in later examples.

Second example – Polyline tool (Fig. 2.24)

This will be a long sequence, but it is typical of a reasonably complex drawing using the **Polyline** tool. In the following sequences, when a prompt line is to be repeated, the prompts in square brackets ([]) will be replaced by **[prompts]** (Fig. 2.24).

40,250   160,250   260,250

260,180

40,120   160,120   260,120

Fig. 2.24 Second example – **Polyline** tool

Left-click the **Polyline** tool icon. The command line shows:

```
Command:_pline Specify start point: 40,250
Current line width is 0
Specify next point or [Arc/Halfwidth/Length/Undo/
  Width]: w (Width)
Specify starting width <0>: 5
Specify ending width <5>: right-click
Specify next point or [Arc/Close/Halfwidth/Length/
  Undo/Width]: 160,250
Specify next point or [prompts]: h (Halfwidth)
Specify starting half-width <2.5>: 1
Specify ending half-width <1>: right-click
Specify next point or [prompts]: 260,250
Specify next point or [prompts]: 260,180
Specify next point or [prompts]: w (Width)
Specify starting width <1>: 10
Specify ending width <10>: right-click
Specify next point or [prompts]: 260,120
Specify next point or [prompts]: h (Halfwidth)
Specify starting half-width <5>: 2
Specify ending half-width <2>: right-click
Specify next point or [prompts]: 160,120
Specify next point or [prompts]: w (Width)
```

```
Specify starting width <4>: 20
Specify ending width <20>: right-click
Specify next point or [prompts]: 40,120
Specify starting width <20>: 5
Specify ending width <5>: right-click
Specify next point or [prompts]: c (Close)
Command:
```

Third example – Polyline tool (Fig. 2.25)

Left-click the **Polyline** tool icon. The command line shows:

```
Command:_pline Specify start point: 50,220
Current line width is 0
[prompts]: w (Width)
Specify starting width <0>: 0.5
Specify ending width <0.5>: right-click
Specify next point or [prompts]: 120,220
Specify next point or [prompts]: a (Arc)
Specify endpoint of arc or [prompts]: s (second pt)
Specify second point on arc: 150,200
Specify end point of arc: 180,220
Specify end point of arc or [prompts]: l (Line)
Specify next point or [prompts]: 250,220
Specify next point or [prompts]: 260,190
Specify next point or [prompts]: a (Arc)
Specify endpoint of arc or [prompts]: s (second pt)
Specify second point on arc: 240,170
Specify end point of arc: 250,160
Specify end point of arc or [prompts]: l (Line)
Specify next point or [prompts]: 250,150
Specify next point or [prompts]: 250,120
```

And so on until the outline Fig. 2.25 is completed.

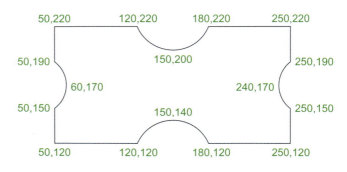

Fig. 2.25 Third example – **Polyline** tool

Fourth example – Polyline tool (Fig. 2.26)

Left-click the **Polyline** tool icon. The command line shows:

```
Command:_pline Specify start point: 80,170
Current line width is 0
Specify next point or [prompts]: w (Width)
Specify starting width <0>: 1
Specify ending width <1>: right-click
Specify next point or [prompts]: a (Arc)
Specify endpoint of arc or [prompts]: s (second pt)
Specify second point on arc: 160,250
Specify end point of arc: 240,170
Specify end point of arc or [prompts]: cl (CLose)
Command:
```

And the circle Fig. 2.26 is formed.

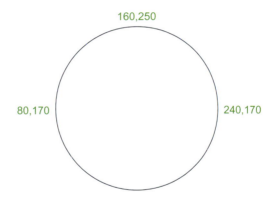

Fig. 2.26 Fourth example – **Polyline** tool

Fifth example – Polyline tool (Fig. 2.27)

Left-click the **Polyline** tool icon. The command line shows:

```
Command:_pline Specify start point: 60,180
Current line width is 0
Specify next point or [prompts]: w (Width)
Specify starting width <0>: 1
Specify ending width <1>: right-click
Specify next point or [prompts]: 190,180
Specify next point or [prompts]: w (Width)
Specify starting width <1>: 20
Specify ending width <20>: 0
Specify next point or [prompts]: 265,180
Specify next point or [prompts]: right-click
Command:
```

CHAPTER 2

And the arrow Fig. 2.27 is formed.

60,180 190,180 265,180
Width=1 Width=20 Width=0

Fig. 2.27 Fifth example – **Polyline** tool

REVISION NOTES

1. **The following terms have been used in this chapter:**
   Left-click – press the left-hand button of the mouse.
   Click – same meaning as *left-click*.
   Double-click – press the left-hand button of the mouse twice.
   Right-click – press the right-hand button of the mouse – has the same result as pressing the Return key of the keyboard.
   Drag – move the cursor on to a feature, and holding down the left-hand button of the mouse pull the object to a new position. Only applies to features such as dialogs and palettes, not to parts of drawings.
   Enter – type the letters of numbers which follow at the keyboard.
   Pick – move the cursor on to an item on screen and press the *left-hand* button of the mouse.
   Return – press the **Enter** key of the keyboard. This key may also marked with a left facing arrow. In most cases (but not always) has the same result as a *right-click*.
   Dialog – a window appearing in the AutoCAD window in which settings may be made.
   Drop-down menu – a menu appearing when one of the names in the menu bar is *clicked*.
   Tooltip – the name of a tool appearing when the cursor is placed over a tool icon.
   Prompts – text appearing in the command window when a tool is selected, which advise the operator as to which operation is required.

2. **Three methods of coordinate entry have been used in this chapter:**
   Absolute method – the coordinates of points on an outline are entered at the command line in response to prompts.
   Relative method – the distances in coordinate units are entered preceded by @ from the last point which has been determined on an outline. Angles, which are measured in a counterclockwise direction, are preceded by <.
   Tracking – the rubber band of the line is dragged in the direction in which the line is to be drawn and its distance in units is entered at the command line followed by a right-click.
   Line and Polyline tools – an outline drawn using the Line tool consists of a number of objects – the number of lines in the outline. An outline drawn using the Polyline is a single object.

Exercises

Methods of constructing answers to the following exercises can be found in the free website:

http://books.elsevier.com/companions/978-0-08-096575-8

1. Using the **Line** tool, construct the rectangle Fig. 2.28.

Fig. 2.28 Exercise 1

2. Construct the outline Fig. 2.29 using the **Line** tool. The coordinate points of each corner of the rectangle will need to be calculated from the lengths of the lines between the corners.

Fig. 2.29 Exercise 2

3. Using the Line tool, construct the outline Fig. 2.30.

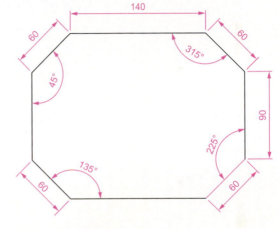

Fig. 2.30 Exercise 3

4. Using the Circle tool, construct the two circles of radius 50 and 30. Then using the Ttr prompt add the circle of radius 25 (Fig. 2.31).

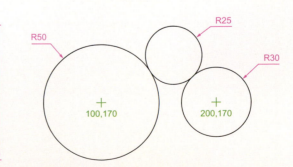

Fig. 2.31 Exercise 4

CHAPTER 2

5. In an acadiso.dwt screen and using the Circle and Line tools, construct the line and circle of radius 40 shown in Fig. 2.32. Then using the Ttr prompt add the circle of radius 25.

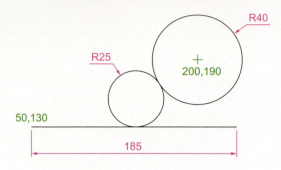

Fig. 2.32 Exercise 5

6. Using the Line tool, construct the two lines at the length and angle as given in Fig. 2.33. Then with the Ttr prompt of the Circle tool, add the circle as shown.

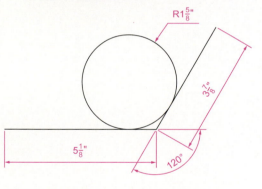

Fig. 2.33 Exercise 6

7. Using the Polyline tool, construct the outline given in Fig. 2.34.

Fig. 2.34 Exercise 7

8. Construct the outline Fig. 2.35 using the Polyline tool.

Fig. 2.35 Exercise 8

9. With the Polyline tool construct the arrows shown in Fig. 2.36.

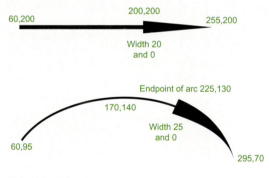

Fig. 2.36 Exercise 9

Draw tools, Object Snap and Dynamic Input

AIMS OF THIS CHAPTER

The aims of this chapter are:

1. To give examples of the use of the **Arc**, **Ellipse**, **Polygon**, **Rectangle**, tools from the **Home/Draw** panel.
2. To give examples of the uses of the **Polyline Edit** (pedit) tool.
3. To introduce the **Object Snap**s (osnap) and their uses.
4. To introduce the **Dynamic Input (DYN)** system and its uses.

Introduction

The majority of tools in AutoCAD 2011 can be called into use by any one of the following six methods:

1. By *clicking* on the tool's icon in the appropriate panel. Fig. 3.1 shows the **Polygon** tool called from the **Home/Draw** panel.

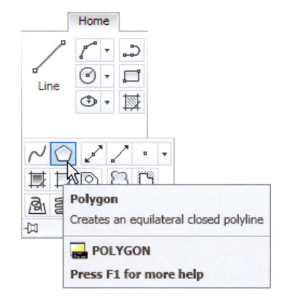

Fig. 3.1 The **Polygon** tool and its tooltip selected from the **Home/Draw** panel

Fig. 3.2 The tool icons in the **Draw** toolbar

2. By *clicking* on a tool icon in a drop-down menu. Fig. 3.2 shows the tool names and icons displayed in the **Draw** drop-down menu. It is necessary to first bring the menu bar to screen with a *click* on **Show Menu Bar** in the *left-click* menu of the **Quick Access** toolbar (Fig. 3.3) if the menu bar is not already on screen.
3. By *entering* an abbreviation for the tool name at the command line. For example, the abbreviation for the **Line** tool is **l**, for the **Polyline** tool it is **pl** and for the **Circle** tool it is **c**.
4. By *entering* the full name of the tool at the command line.
5. By making use of the **Dynamic Input** method of construction.
6. If working in the **AutoCAD Classic** workspace by selection of tools from toolbars.

In practice operators constructing drawings in AutoCAD 2011 may well use a combination of these six methods.

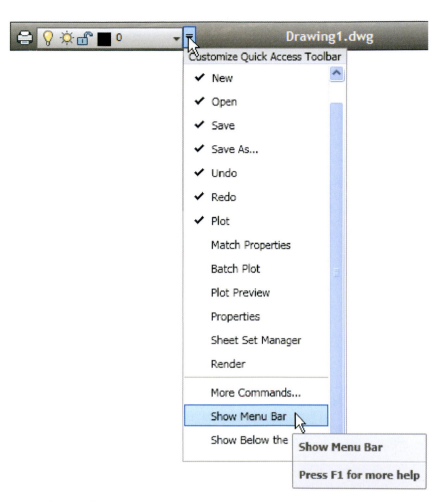

Fig. 3.3 Selecting **Show Menu Bar** from the *left-click* menu in the **Quick Access** toolbar

The Arc tool

In AutoCAD 2011, arcs can be constructed using any three of the following characteristics of an arc – its **Start** point, a point on the arc (**Second** point), its **Center**, its **End**, its **Radius**, the **Length** of the arc, the **Direction** in which the arc is to be constructed, the **Angle** between lines of the arc.

These characteristics are shown in the menu appearing with a *click* on the arrow to the right of the **Arc** tool icon in the **Home/Draw** panel (Fig. 3.4).

To call the **Arc** tool *click* on the flyout of its tool icon in the **Home/Draw** panel, *click* on **Arc** in the **Draw** drop-down menu or *enter* **a** or **arc** at the command line. In the following examples, initials of prompts will be shown instead of selection from the menu as shown in Fig. 3.5.

CHAPTER 3

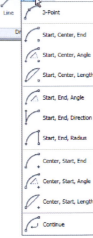

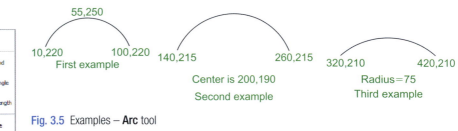

Fig. 3.5 Examples – **Arc** tool

Fig. 3.4 The **Arc** tool flyout in the **Home/Draw** panel

First example – Arc tool (Fig. 3.5)

Left-click the **Arc** tool icon. The command line shows:

```
Command:_arc Specify start point of arc or
   [Center]: 100,220
Specify second point of arc or [Center/End]:
   55,250
Specify end point of arc: 10,220
Command:
```

Second example – Arc tool (Fig. 3.5)

```
Command:right-click brings back the Arc sequence
ARC Specify start point of arc or [Center]: c
   (Center)
Specify center point of arc: 200,190
Specify start point of arc: 260,215
Specify end point of arc or [Angle/chord Length]:
   140,215
Command:
```

Third example – Arc tool (Fig. 3.5)

```
Command:right-click brings back the Arc sequence
ARC Specify start point of arc or [Center]:
   420,210
Specify second point of arc or [Center/End]:
   e (End)
Specify end point of arc: 320,210
Specify center point of arc or [Angle/Direction/
   Radius]: r (Radius)
Specify radius of arc: 75
Command:
```

The Ellipse tool

Ellipses can be regarded as what is seen when a circle is viewed from directly in front of the circle and the circle rotated through an angle about its horizontal diameter. Ellipses are measured in terms of two axes – a **major axis** and a **minor axis**, the major axis being the diameter of the circle and the minor axis being the height of the ellipse after the circle has been rotated through an angle (Fig. 3.6).

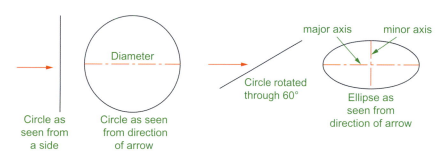

Fig. 3.6 An ellipse can be regarded as viewing a rotated circle

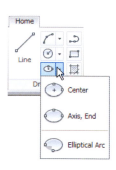

Fig. 3.7 The **Ellipse** tool icon flyout in the **Home/Draw** panel

To call the **Ellipse** tool, *click* on its tool icon in the **Home/Draw** panel (Fig. 3.7), *click* its name in the **Draw** drop-down menu or *enter* **a** or **arc** at the command line.

First example – Ellipse (Fig. 3.8)

Left-click the **Ellipse** tool icon. The command line shows:

```
Command:_ellipse
Specify axis endpoint of elliptical arc or
   [Center]: 30,190
Specify other endpoint of axis: 150,190
Specify distance to other axis or [Rotation] 25
Command:
```

Second example – Ellipse (Fig. 3.8)

In this second example, the coordinates of the centre of the ellipse (the point where the two axes intersect) are *entered*, followed by *entering* coordinates for the end of the major axis, followed by *entering* the units for the end of the minor axis.

```
Command:right-click
ELLIPSE
```

```
Specify axis endpoint of elliptical arc or
   [Center]: c
Specify center of ellipse: 260,190
Specify endpoint of axis: 205,190
Specify distance to other axis or
[Rotation]: 30
Command:
```

Third example – Ellipse (Fig. 3.8)

In this third example, after setting the positions of the ends of the major axis, the angle of rotation of the circle from which an ellipse can be obtained is *entered* (Fig. 3.8).

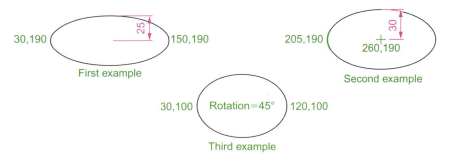

Fig. 3.8 Examples – **Ellipse**

```
Command: right-click
ELLIPSE
Specify axis endpoint of elliptical arc or
   [Center]: 30,100
Specify other endpoint of axis: 120,100
Specify distance to other axis or [Rotation]:
   r (Rotation)
Specify rotation around major axis: 45
Command:
```

Saving drawings

Before going further it is as well to know how to save the drawings constructed when answering examples and exercises in this book. When a drawing has been constructed, *left-click* on **Save As** in the menu appearing with a *left-click* on the AutoCAD icon at the top left-hand corner of the window (Fig. 3.9). The **Save Drawing As** dialog appears (Fig. 3.10).

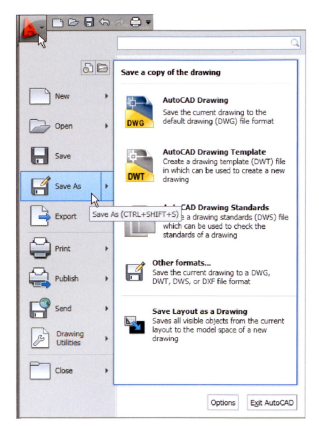

Fig. 3.9 Selecting **Save As** from the **Quick Access** menu

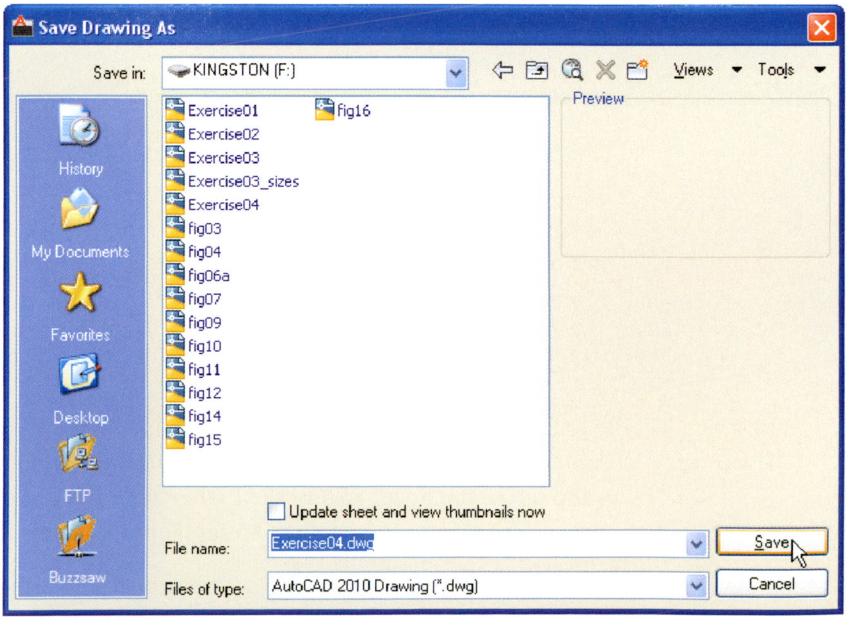

Fig. 3.10 The **Save Drawing As** dialog

Unless you are the only person using the computer on which the drawing has been constructed, it is best to save work to a USB memory stick or other form of temporary saving device. To save a drawing to a USB memory stick:

1. Place a memory stick in a **USB** drive.
2. In the **Save in:** field of the dialog, *click* the arrow to the right of the field and from the popup list select **KINGSTON [F:]** (the name of my **USB** drive and stick).
3. In the **File name:** field type a name. The file name extension **.dwg** does not need to be typed – it will be added to the file name.
4. *Left-click* the **Save** button of the dialog. The drawing will be saved with the file name extension **.dwg** – the AutoCAD file name extension (Fig. 3.10).

Snap

In previous chapters, several methods of constructing accurate drawings have been described – using **Snap**, absolute coordinate entry, relative coordinate entry and tracking. Other methods of ensuring accuracy between parts of constructions are by making use of **Object Snaps** (**Osnaps**).

Snap Mode, **Grid Display** and **Object Snaps** can be toggled on/off from the buttons in the status bar or by pressing the keys, **F9** (**Snap Mode**), **F7** (**Grid Display**) and **F3** (**Object Snap**).

Object Snaps (Osnaps)

Object Snaps allow objects to be added to a drawing at precise positions in relation to other objects already on screen. With Object Snaps, objects can be added to the end points, midpoints, to intersections of objects, to centres and/or quadrants of circles and so on. Object Snaps also override snap points even when snap is set on.

To set **Object Snaps** – at the command line:

```
Command: enter os
```

And the **Drafting Settings** dialog appears (Fig. 3.11). *Click* the **Object Snap** tab in the upper part of the dialog and *click* the check boxes to the right of the Object Snap names to set them on (or off in on).

When Object Snaps are set **ON**, as outlines are constructed using Object Snap icons and their tooltips appear as indicated in Fig. 3.12.

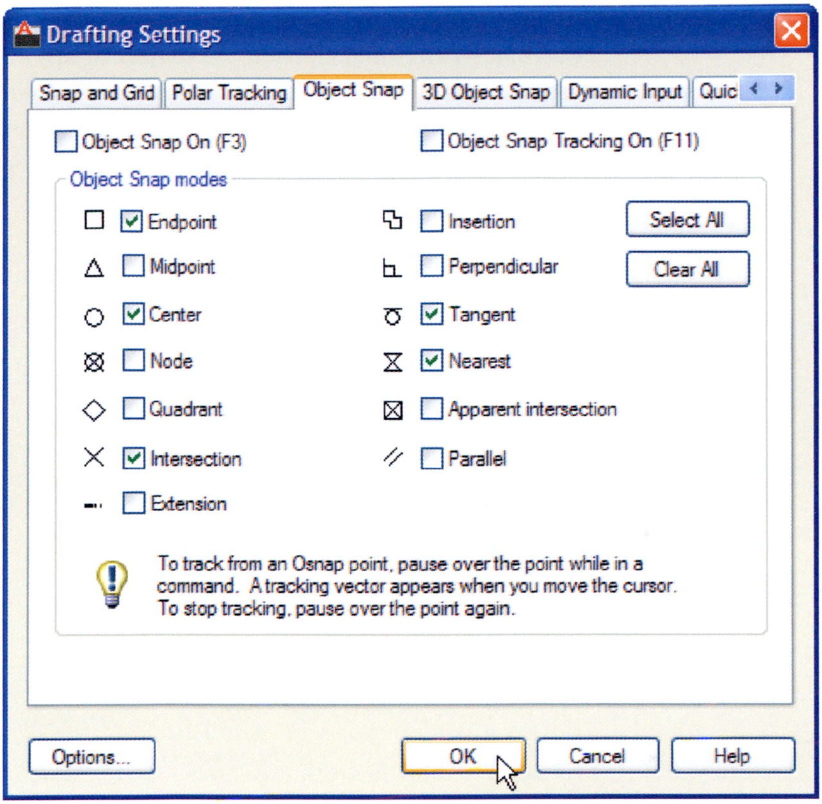

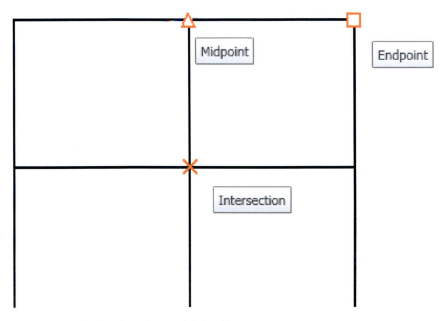

Fig. 3.11 The **Drafting Settings** dialog with some of the **Object Snaps** set on

Fig. 3.12 Three **Object Snap** icons and their tooltips

CHAPTER 3

It is sometimes advisable not to have **Object Snaps** set on in the **Drafting Settings** dialog, but to set **Object Snap** off and use Object Snap abbreviations at the command line when using tools. The following examples show the use of some of these abbreviations. **Object Snaps** can be toggled on/off by pressing the **F3** key of the keyboard.

First example – Object Snap (Fig. 3.13)

Call the **Polyline** tool:

```
Command:_pline
Specify start point: 50,230
[prompts]: w (Width)
Specify starting width: 1
Specify ending width <1>: right-click
Specify next point: 260,230
Specify next point: right-click
Command: right-click
PLINE
Specify start point: pick the right-hand end of
  the pline
Specify next point: 50,120
Specify next point: right-click
Command: right-click
PLINE
Specify start point: pick near the middle of first
  pline
Specify next point: 155,120
Specify next point: right-click
Command: right-click
PLINE
Specify start point: pick the plines at their
  intersection
Specify start point: right-click
Command:
```

The result is shown in Fig. 3.13. In this illustration the Object Snap tooltips are shown as they appear when each object is added to the outline.

Second example – Object Snap abbreviations (Fig. 3.14)

Call the **Circle** tool:

```
Command:_circle
Specify center point for circle: 180,170
```

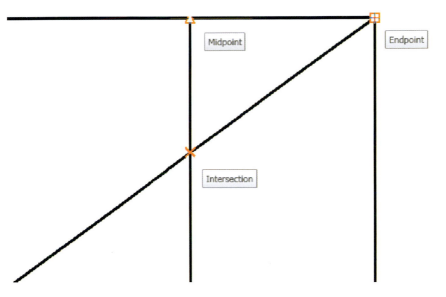

Fig. 3.13 First example – **Osnaps**

```
Specify radius of circle: 60
Command: enter l (Line) right-click
Specify first point: enter qua right-click
of pick near the upper quadrant of the circle
Specify next point: enter cen right-click
of pick near the centre of the circle
Specify next point: enter qua right-click
of pick near right-hand side of circle
Specify next point: right-click
Command:
```

Notes

With Object Snaps off, the following abbreviations can be used:

end – endpoint;
mid – midpoint;
int – intersection;
cen – centre;
qua – quadrant;

CHAPTER 3

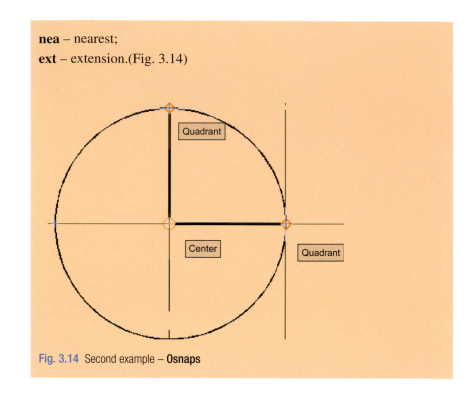

nea – nearest;

ext – extension.(Fig. 3.14)

Quadrant

Center

Quadrant

Fig. 3.14 Second example – **Osnaps**

Dynamic Input (DYN)

When **Dynamic Input** is set on by either pressing the **F12** key or with a *click* on the **Dynamic Input** button in the status bar, dimensions, coordinate positions and commands appear as tips when no tool is in action (Fig. 3.15).

With a tool in action, as the cursor hairs are moved in response to movement of the mouse, **Dynamic Input** tips showing the coordinate figures for the point

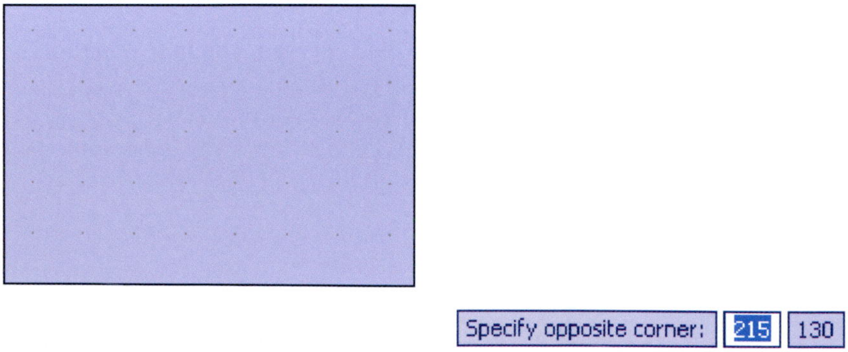

Specify opposite corner: 215   130

Fig. 3.15 The **DYN** tips appearing when no tool is in action and the cursor is moved

of the cursor hairs will show (Fig. 3.16), together with other details. To see the drop-down menu giving the prompts available with **Dynamic Input** press the down key of the keyboard and *click* the prompt to be used. Fig. 3.16 shows the **Arc** prompt as being the next to be used when the **Polyline** tool is in use.

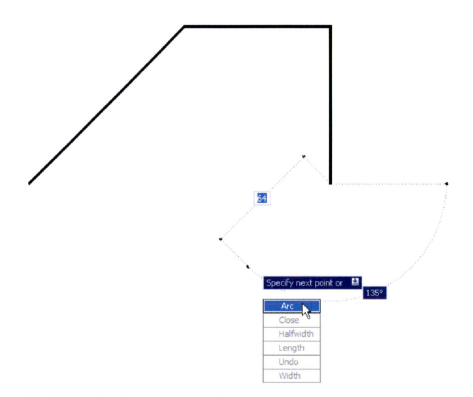

Fig. 3.16 Coordinate tips when **DYN** is in action

Notes on the use of Dynamic Input

Although **Dynamic Input** can be used in any of the AutoCAD 2011 workspaces, some operators may prefer a larger working area. To achieve this a *click* on the **Clean Screen** icon in the bottom right-hand corner of the AutoCAD 2011 window produces an uncluttered workspace area. The command palette can be cleared from screen by *entering* **commandlinehide** at the command line. To bring it back press the keys **Ctrl+9**. These two operations produce a screen showing only title and status bars (Fig. 3.17). Some operators may well prefer working in such a larger than normal workspace.

Dynamic Input settings are made in the **Dynamic Input** sub-dialog of the **Drafting Settings** dialog (Fig. 3.18), brought to screen by *entering* **os** (or **ds**) at the command line.

CHAPTER 3

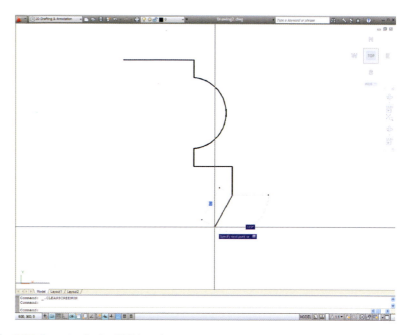

Fig. 3.17 Example of using **DYN** in a clear screen

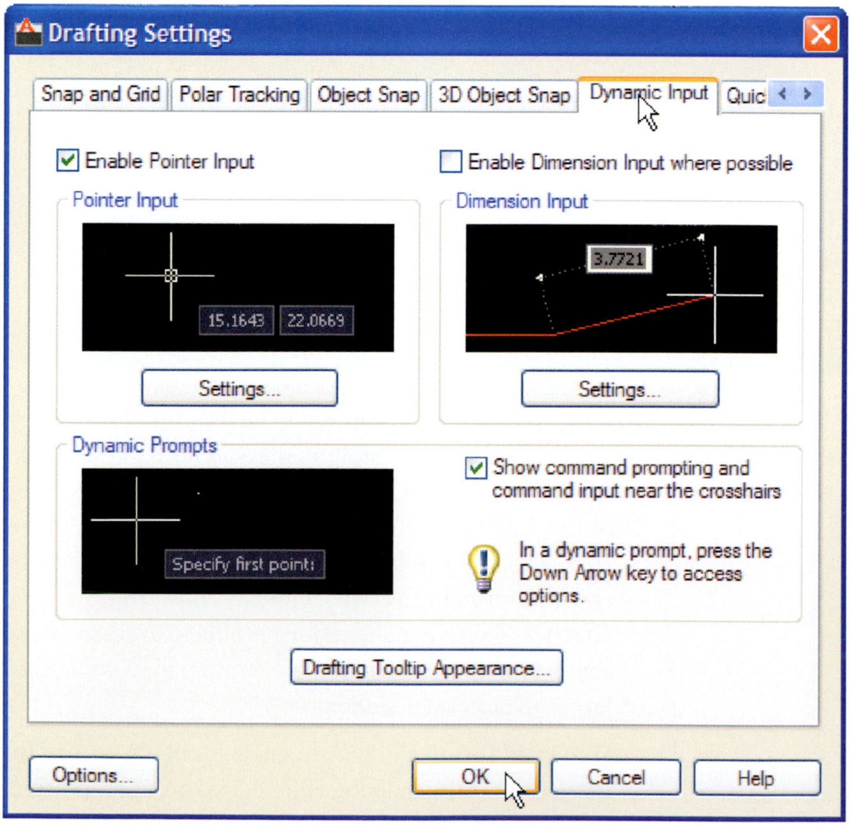

Fig. 3.18 Settings for **DYN** can be made in the **Drafting Settings** dialog

When **Dynamic Input** is in action, tools can be called by using any of the methods described on page 50.

1. By *entering* the name of the tool at the command line.
2. By *entering* the abbreviation for a tool name at the command line.
3. By selecting the tool's icon from a panel.
4. By selecting the tool's name from a drop-down menu.

When **Dynamic Input** is active and a tool is called, command prompts appear in a tooltip at the cursor position. Fig. 3.19 shows the tooltip appearing at the cursor position when the **Line** tool icon in the **Home/Draw** panel is *clicked*.

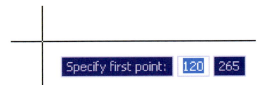

Fig. 3.19 The prompt appearing on screen when the **Line** tool is selected

To commence drawing a line, either move the cursor under mouse control to the desired coordinate point and *left-click* as in Fig. 3.20, or *enter* the required *x,y* coordinates at the keyboard (Fig. 3.21) and *left-click*. To continue drawing with **Line** *drag* the cursor to a new position and either *left-click* at the position when the coordinates appear as required (Fig. 3.21), or *enter* a required length at the keyboard, which appears in the length box followed by a *left-click* (Fig. 3.22).

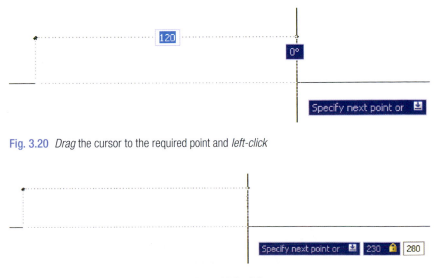

Fig. 3.20 *Drag* the cursor to the required point and *left-click*

Fig. 3.21 *Enter* coordinates for the next point and *left-click*

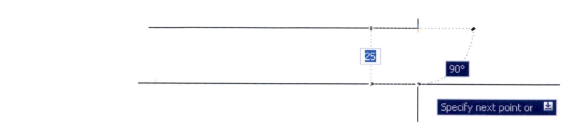

Fig. 3.22 *Enter* length at keyboard and *right-click*

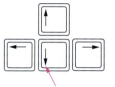

The down key of the keyboard

Fig. 3.23 The **down** key of the keyboard

When using **Dynamic Input** the selection of a prompt can be made by pressing the **down** key of the keyboard (Fig. 3.23) which causes a popup menu to appear. A *click* on the required prompt in such a popup menu will make that prompt active.

Fig. 3.24 Selecting **Polyline** from the **Home/Draw** panel

Dynamic Input – first example – Polyline

1. Select **Polyline** from the **Home/Draw** panel (Fig. 3.24).
2. To start the construction *click* at any point on screen. The prompt for the polyline appears with the coordinates of the selected point showing. *Left-click* to start the drawing (Fig. 3.25).

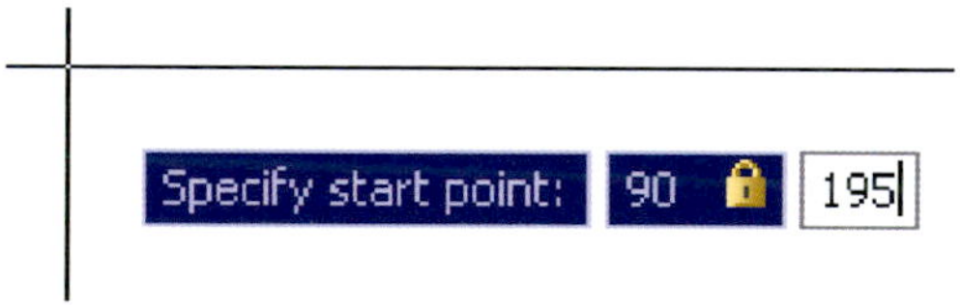

Fig. 3.25 **Dynamic Input** – first example – **Polyline** – the first prompt

3. Move the cursor and press the down key of the keyboard. A popup menu appears from which a prompt selection can be made. In the menu *click* **Width** (Fig. 3.26).
4. Another prompt field appears. At the keyboard *enter* the required width and *right-click*. Then *left-click* and *enter* **ending width** or *right-click* if the **ending width** is the same as the **starting width** (Fig. 3.27).
5. *Drag* the cursor to the right until the dimension shows the required horizontal length and *left-click* (Fig. 3.28).
6. *Drag* the cursor down until the vertical distance shows and *left-click* (Fig. 3.29).
7. *Drag* the cursor to the left until the required horizontal distance is showing and *right-click* (Fig. 3.30).
8. Press the down key of the keyboard and *click* **Close** in the menu (Fig. 3.31). The rectangle completes.

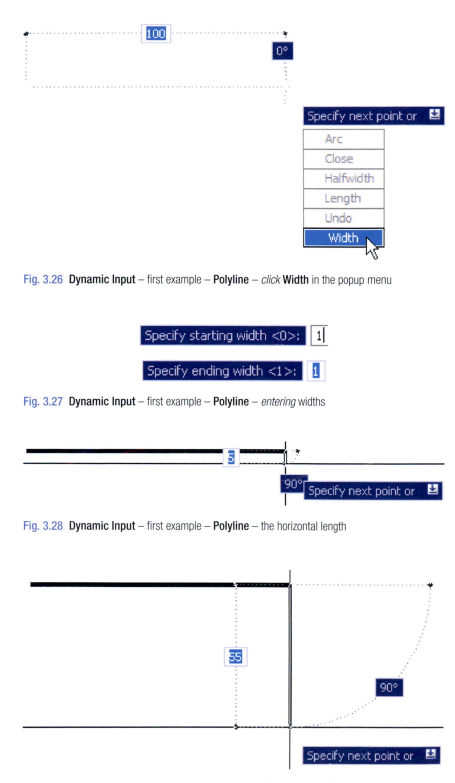

CHAPTER 3

Fig. 3.26 **Dynamic Input** – first example – **Polyline** – *click* **Width** in the popup menu

Fig. 3.27 **Dynamic Input** – first example – **Polyline** – *entering* widths

Fig. 3.28 **Dynamic Input** – first example – **Polyline** – the horizontal length

Fig. 3.29 **Dynamic Input** – first example – **Polyline** – the vertical height

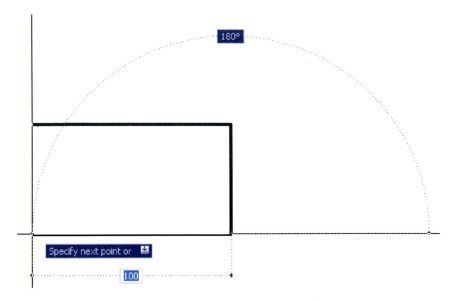

Fig. 3.30 **Dynamic Input** – first example – **Polyline** – the horizontal distance

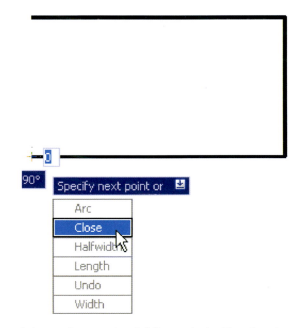

Fig. 3.31 **Dynamic Input** – first example – **Polyline** – selecting **Close** from the popup menu

Fig. 3.32 shows the completed drawing.

DYN – second example – Zoom

1. *Enter* **Zoom** or **z** at the command line. The first **Zoom** prompt appears (Fig. 3.33).

Fig. 3.32 Dynamic Input – first example – **Polyline**

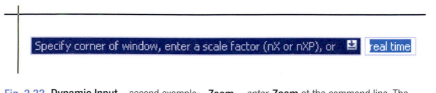

Fig. 3.33 Dynamic Input – second example – **Zoom** – *enter* **Zoom** at the command line. The prompts which then appear

2. *Right-click* and press the down button of the keyboard. The popup list (Fig. 3.34) appears from which a **Zoom** prompt can be selected.
3. Carry on using the **Zoom** tool as described in Chapter 4.

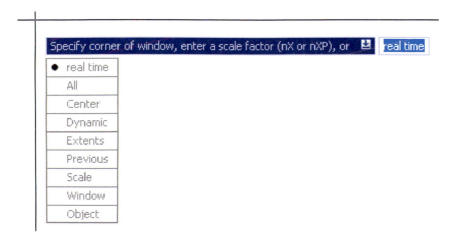

Fig. 3.34 Dynamic Input – second example – **Zoom** – the popup menu appearing with a *right-click* and pressing the **down** keyboard button

DYN – third example – dimensioning

When using **DYN**, tools can equally as well be selected from a panel. Fig. 3.35 shows the **Linear** tool from the **Home/Annotation** panel selected when dimensioning a drawing.

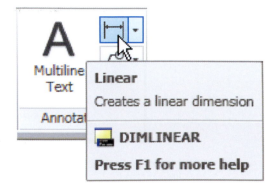

Fig. 3.35 Selecting **Linear** from the **Home/Annotation** panel

A prompt appears asking for the first point. Move the cursor to the second point, another prompt appears (Fig. 3.36). Press the down button of the keyboard and the popup list (Fig. 3.36) appears from which a selection can be made.

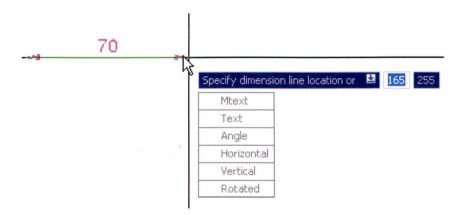
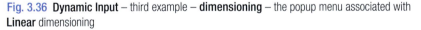

Fig. 3.36 **Dynamic Input** – third example – **dimensioning** – the popup menu associated with **Linear** dimensioning

The **Dynamic Input** method of constructing 2D drawings can equally as well be used when constructing 3D solid models drawings (see Chapter 12 onwards).

Why use Dynamic Input?

Some operators may prefer constructing drawings without having to make entries at the command line in response to tool prompts. By using **DYN** drawings, whether in 2D or in 3D format, can be constructed purely from operating and moving the mouse, *entering* coordinates at the command line and pressing the down key of the keyboard when necessary.

Examples of using other Draw tools

Polygon tool (Fig. 3.37)

Call the **Polygon** tool – either with a *click* on its tool icon in the **Home/ Draw** panel (Fig. 3.1, page 69), from the **Draw** drop-down menu, or by *entering* **pol** or **polygon** at the command line. No matter how the tool is called, the command line shows:

```
Command:_polygon Enter number of sides <4>: 6
Specify center of polygon or [Edge]: 60,210
Enter an option [Inscribed in circle/Circumscribed
  about circle] <I>: right-click (accept Inscribed)
Specify radius of circle: 60
Command:
```

1. In the same manner construct a **5**-sided polygon of centre **200,210** and of radius **60**.
2. Then, construct an **8**-sided polygon of centre **330,210** and radius **60**.
3. Repeat to construct a **9**-sided polygon circumscribed about a circle of radius **60** and centre **60,80**.
4. Construct yet another polygon with **10** sides of radius **60** and of centre **200,80**.
5. Finally another polygon circumscribing a circle of radius **60**, of centre **330,80** and sides **12**.

The result is shown in Fig. 3.37.

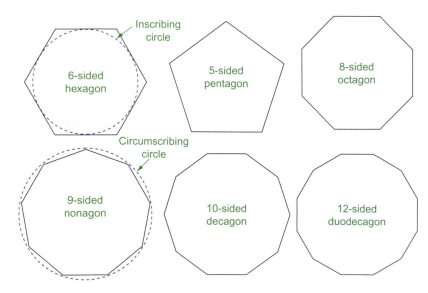

Inscribing circle

6-sided hexagon

5-sided pentagon

8-sided octagon

Circumscribing circle

9-sided nonagon

10-sided decagon

12-sided duodecagon

Fig. 3.37 First example – **Polygon** tool

Rectangle tool – first example (Fig. 3.39)

Call the **Rectangle** tool – either with a *click* on its tool icon in the **Home/Draw** panel (Fig. 3.38) by *entering* **rec** or **rectangle** at the command line. The tool can be also called from the **Draw** drop-down menu. The command line shows (Fig. 3.39):

Fig. 3.38 The **Rectangle** tool from the **Home/Draw** panel

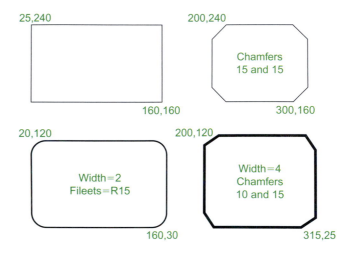

Fig. 3.39 Examples – **Rectangle** tool

```
Command:_rectang
Specify first corner point or [Chamfer/
Elevation/Fillet/Thickness/Width]: 25,240
Specify other corner point or [Area/Dimensions/
  Rotation]: 160,160
Command:
```

Rectangle tool – second example (Fig. 3.39)

```
Command:_rectang
[prompts]: c (Chamfer)
Specify first chamfer distance for rectangles
  <0>: 15
Specify first chamfer distance for rectangles <15>:
  right-click
Specify first corner point: 200,240
Specify other corner point: 300,160
Command:
```

Rectangle tool – third example (Fig. 3.39)

```
Command: _rectang
Specify first corner point or [Chamfer/Elevation/
  Fillet/Thickness/Width]: f (Fillet)
Specify fillet radius for rectangles <0>: 15
Specify first corner point or [Chamfer/Elevation/
  Fillet/Thickness/Width]: w (Width)
Specify line width for rectangles <0>: 1
Specify first corner point or [Chamfer/Elevation/
  Fillet/Thickness/Width]: 20,120
Specify other corner point or [Area/Dimensions/
  Rotation]: 160,30
Command:
```

Rectangle – fourth example (Fig. 3.39)

```
Command:_rectang
Specify first corner point or [Chamfer/Elevation/
  Fillet/Thickness/Width]: w (Width)
Specify line width for rectangles <0>: 4
Specify first corner point or [Chamfer/Elevation/
  Fillet/Thickness/Width]: c (Chamfer)
Specify first chamfer distance for rectangles <0>: 15
Specify second chamfer distance for rectangles
  <15>: right-click
Specify first corner point: 200,120
Specify other corner point: 315,25
Command:
```

The Polyline Edit tool

The **Polyline Edit** tool is a valuable tool for the editing of polylines.

First example – Polyline Edit (Fig. 3.42)

1. With the **Polyline** tool construct the outlines **1** to **6** of Fig. 3.40.
2. Call the **Edit Polyline** tool either from the **Home/Modify** panel (Fig. 3.41) or from the **Modify** drop-down menu, or by *entering* **pe** or **pedit** at the command line, which then shows:

```
Command: enter pe
PEDIT Select polyline or [Multiple]: pick pline 2
```

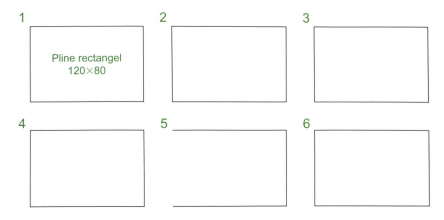

Fig. 3.40 Examples – **Edit Polyline** – the plines to be edited

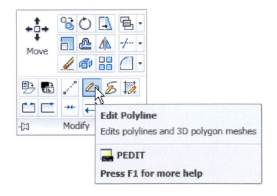

Fig. 3.41 Calling **Edit Polyline** from the **Home/Modify** panel

```
Enter an option [Open/Join/Width/Edit vertex/Fit/
  Spline/Decurve/Ltype gen/Reverse/Undo]:
  w (Width)
Specify new width for all segments: 2
Enter an option [Open/Join/Width/Edit vertex/Fit/
  Spline/Decurve/Ltype gen/Reverse/Undo]: right-
  click
Command:
```

3. Repeat with pline **3** and pedit to Width = **10**.
4. Repeat with line **4** and *enter* **s** (Spline) in response to the prompt line:

```
Enter an option [Open/Join/Width/Edit vertex/Fit/
  Spline/Decurve/Ltype gen/Reverse/Undo]: enter s
```

5. Repeat with pline **5** and *enter* **j** in response to the prompt line:

```
Enter an option [Open/Join/Width/Edit vertex/Fit/
   Spline/Decurve/Ltype gen/Undo]: enter j
```

The result is shown in pline **6**.

The resulting examples are shown in Fig. 3.42.

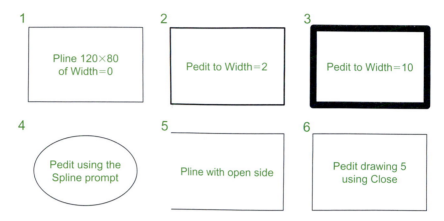

1 Pline 120×80 of Width=0

2 Pedit to Width=2

3 Pedit to Width=10

4 Pedit using the Spline prompt

5 Pline with open side

6 Pedit drawing 5 using Close

Fig. 3.42 Examples – **Polyline Edit**

Example – Multiple Polyline Edit (Fig. 3.43)

1. With the **Polyline** tool construct the left-hand outlines of Fig. 3.43.
2. Call the **Edit Polyline** tool. The command line shows:

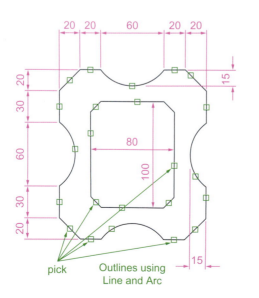

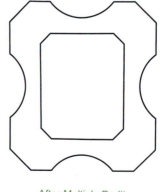

pick   Outlines using Line and Arc

After Multiple Pedit to Width=2

Fig. 3.43 Example – **Multiple Polyline Edit**

```
Command: enter pe
PEDIT Select polyline or [Multiple]: m (Multiple)
Select objects: pick any one of the lines or arcs
  of the left-hand outlines of Fig. 6.16 1 found
Select objects: pick another line or arc 1 found 2
  total
Continue selecting lines and arcs as shown by the
  pick boxes of the left-hand drawing of Fig. 3.45
  until the command line shows:
Select objects: pick another line or arc 1 found
  24 total
Select objects: right-click
[prompts]: w (Width)
Specify new width for all segments: 1.5
Convert Arcs, Lines and Splines to polylines [Yes/
  No]? <Y>: right-click
[prompts]: right-click
Command:
```

The result is shown in the right-hand drawing of Fig. 3.43.

Transparent commands

When any tool is in operation it can be interrupted by prefixing the interrupting command with an apostrophe ('). This is particularly useful when wishing to zoom when constructing a drawing (see page 82). As an example when the **Line** tool is being used:

```
Command:_line
Specify first point: 100,120
Specify next point: 190,120
Specify next point: enter 'z (Zoom)
>> Specify corner of window or [prompts]: pick
>>>> Specify opposite corner: pick
Resuming line command.
Specify next point:
```

And so on. The transparent command method can be used with any tool.

The set variable PELLIPSE

Many of the operations performed in AutoCAD are carried out under settings of **SET VARIABLES**. Some of the numerous set variables available in AutoCAD 2011 will be described in later pages. The variable **PELLIPSE** controls whether ellipses are drawn as splines or as polylines. It is set as follows:

```
Command: enter pellipse right-click
Enter new value for PELLIPSE <0>: enter 1 right-
  click
Command:
```

And now when ellipses are drawn they are plines. If the variable is set to **0**, the ellipses will be splines. The value of changing ellipses to plines is that they can then be edited using the **Polyline Edit** tool.

REVISION NOTES

The following terms have been used in this chapter:

Field – a part of a window or of a dialog in which numbers or letters are *entered* or which can be read.

Popup list – a list brought in screen with a *click* on the arrow often found at the right-hand end of a field.

Object – a part of a drawing which can be treated as a single object. For example, a line constructed with the Line tool is an object, a rectangle constructed with the Polyline tool is an object and an arc constructed with the Arc tool is an object. It will be seen in a later chapter (Chapter 9) that several objects can be formed into a single object.

Ribbon palettes – when working in either of the 2D Drafting and Annotation or of the 3D Modeling workspace, tool icons are held in panels in the Ribbon.

Command line – a line in the command palette which commences with the word Command.

Snap Mode, Grid Display and Object Snap can be toggled with *clicks* on their respective buttons in the status bar. These functions can also be set with function keys: Snap Mode – F9, Grid Display – F7 and Object Snap – F3.

Object Snaps ensure accurate positioning of objects in drawings.

Object Snap abbreviations can be used at the command line rather than setting in ON in the Drafting Settings dialog.

Dynamic input allows constructions in any of the three AutoCAD 2011 workspaces or in a full screen workspace, without having to use the command palette for *entering* the initials of command line prompts.

Notes

There are two types of tooltip. When the cursor under mouse control is paced over a tool icon, the first (a smaller) tooltip is seen. If the cursor is held in position for a short time the second (a larger) tooltip is seen. Settings for the tooltip may be made in the **Options** dialog.

Polygons constructed with the **Polygon** tool are regular polygons – the edges of the polygons are all the same length and the angles are of the same degrees.

Polygons constructed with the **Polygon** tool are plines, so can be edited by using the **Edit Polyline** tool.

The easiest method of calling the **Edit Polyline** tool is to *enter* **pe** at the command line.

The **Multiple** prompt of the **pedit** tool saves considerable time when editing a number of objects in a drawing.

Transparent commands can be used to interrupt tools in operation by preceding the interrupting tool name with an apostrophe (').

Ellipses drawn when the variable **PELLIPSE** is set to **0** are splines; when **PELLIPSE** is set to **1**, ellipses are polylines. When ellipses are in polyline form they can be modified using the **pedit** tool.

CHAPTER 3

Exercises

Methods of constructing answers to the following exercises can be found in the free website:

http://books.elsevier.com/companions/978-0-08-096575-8

1. Using the **Line** and **Arc** tools, construct the outline given in Fig. 3.44.

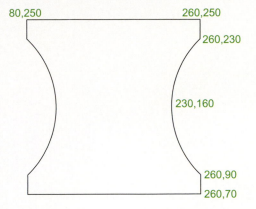

Fig. 3.44 Exercise 1

2. With the **Line** and **Arc** tools, construct the outline Fig. 3.45.

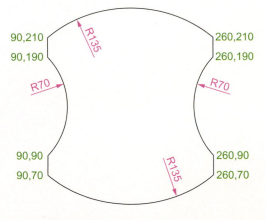

Fig. 3.45 Exercise 2

3. Using the **Ellipse** and **Arc** tools, construct the drawing Fig. 3.46.

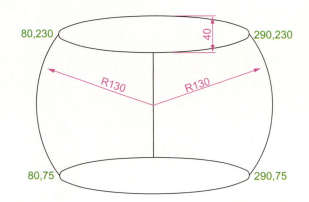

Fig. 3.46 Exercise 3

4. With the **Line**, **Circle** and **Ellipse** tools, construct Fig. 3.47.

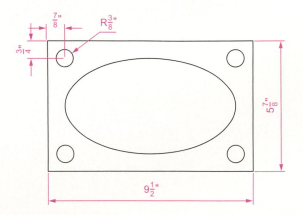

Fig. 3.47 Exercise 4

5. With the **Ellipse** tool, construct the drawing Fig. 3.48.

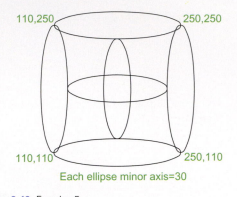

Each ellipse minor axis=30

Fig. 3.48 Exercise 5

6. Fig. 3.49 shows a rectangle in the form of a square with hexagons along each edge. Using the **Dimensions** prompt of the **Rectangle** tool, construct the square. Then, using the **Edge** prompt of the **Polygon** tool, add the four hexagons. Use the **Object Snap endpoint** to ensure the polygons are in their exact positions.

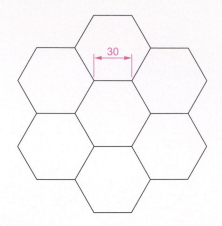

Fig. 3.49 Exercise 6

7. Fig. 3.50 shows seven hexagons with edges touching. Construct the inner hexagon using the **Polygon** tool, then with the aid of the **Edge** prompt of the tool, add the other six hexagons.

Fig. 3.50 Exercise 7

8. Fig. 3.51 was constructed using only the **Rectangle** tool. Make an exact copy of the drawing using only the **Rectangle** tool.

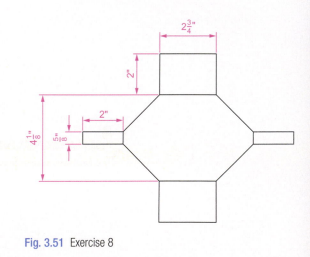

Fig. 3.51 Exercise 8

9. Construct the drawing Fig. 3.52 using the **Line** and **Arc** tools. Then, with the aid of the **Multiple** prompt of the **Edit Polyline** tool, change the outlines into plines of **Width=1**.

10. Construct Fig. 3.53 using the **Line** and **Arc** tools. Then change all widths of lines and arcs to a width of **2** with **Polyline Edit**.

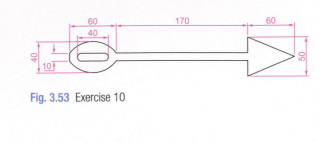

Fig. 3.53 Exercise 10

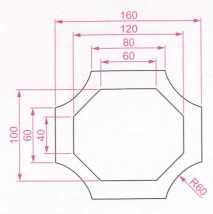

Fig. 3.52 Exercise 9

11. Construct Fig. 3.54 using the **Rectangle**, **Line** and **Edit Polyline** tools.

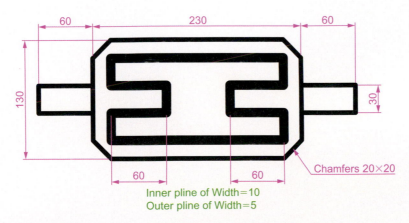

Inner pline of Width=10
Outer pline of Width=5

Fig. 3.54 Exercise 11

Zoom, Pan and templates

AIMS OF THIS CHAPTER

The aims of this chapter are:

1. To demonstrate the value of the **Zoom** tools.
2. To introduce the **Pan** tool.
3. To describe the value of using the **Aerial View** window in conjunction with the **Zoom** and **Pan** tools.
4. To update the **acadiso.dwt** template.
5. To describe the construction and saving of drawing templates.

Introduction

The use of the **Zoom** tools allows not only the close inspection of the most minute areas of a drawing in the AutoCAD 2011 drawing area, but also the accurate construction of very small details in a drawing.

The **Zoom** tools can be called by selection from the **View/Navigate** panel or from the **View** drop-down menu (Fig. 4.1). However by far the easiest and quickest method of calling the **Zoom** is to *enter* **z** at the command line as follows:

```
Command: enter z right-click
ZOOM Specify corner of window, enter a scale factor
  (nX or nXP) or [All/Center/Dynamic/Extents/
  Previous/Scale/Window/Object] <real time>:
```

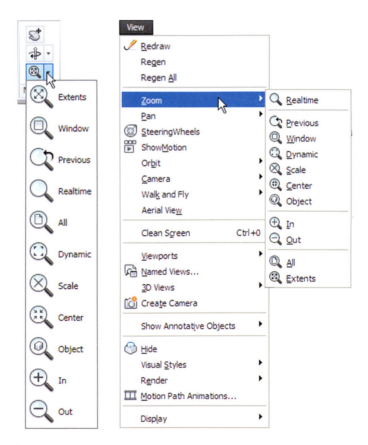

Fig. 4.1 Calling **Zoom** – from the **Zoom/Navigate** panel or from the **View** drop-down menu

This allows the different zooms:

Realtime – selects parts of a drawing within a window.
All – the screen reverts to the limits of the template.

Center – the drawing centres itself around a *picked* point.

Dynamic – a broken line surrounds the drawing which can be changed in size and repositioned to part of the drawing.

Extents – the drawing fills the AutoCAD drawing area.

Previous – the screen reverts to its previous zoom.

Scale – entering a number or a decimal fraction scales the drawing.

Window – the parts of the drawing within a *picked* window appears on screen. The effect is the same as using **real time**.

Object – *pick* any object on screen and the object zooms.

The operator will probably be using **Realtime**, **Window** and **Previous** zooms most frequently.

Figs 4.2–4.4 show a drawing which has been constructed, a **Zoom Window** of part of the drawing allowing it to be checked for accuracy and a **Zoom Extents**, respectively.

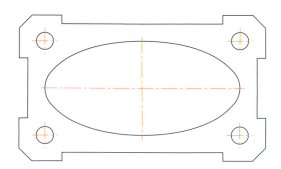

Fig. 4.2 Drawing to be acted upon by the **Zoom** tool

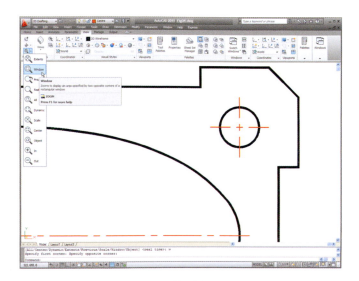

Fig. 4.3 **Zoom Window** of part of the drawing Fig. 4.2

CHAPTER 4

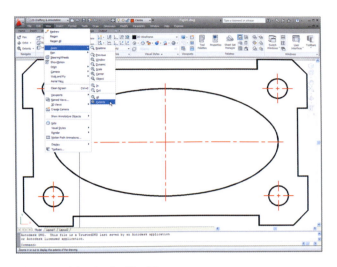

Fig. 4.4 **Zoom Extents** of the drawing Fig. 4.2

It will be found that the **Zoom** tools are among those most frequently used when working in AutoCAD 2011.

The Aerial View window

Enter **dsviewer** at the command line and the **Aerial View** window appears – usually in the bottom right-hand corner of the AutoCAD 2011

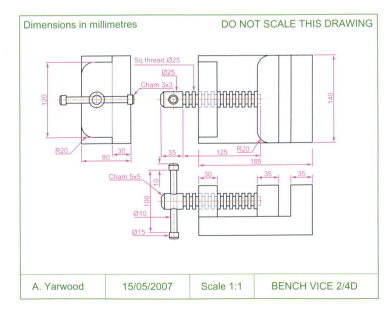

Fig. 4.5 The drawing used to illustrate Figs 4.6 and 4.7

window. The **Aerial View** window shows the whole of a drawing, even if larger that the limits. The **Aerial View** window is of value when dealing with large drawings – it allows that part of the window on screen to be shown in relation to the whole of the drawing. Fig. 4.5 is a three-view orthographic projection of a small bench vice.

Fig. 4.6 shows a **Zoom Window** of the drawing Fig. 4.5 including the **Aerial View Window**. The area of the drawing within the **Zoom** window in the drawing area is bounded by a thick green line in the **Aerial View** window.

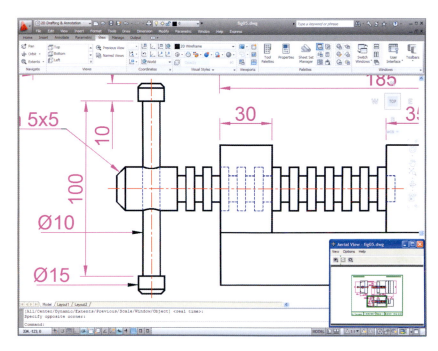

Fig. 4.6 **Zoom Window** of the drawing Fig. 4.5 with its surrounding zoom rectangle showing in the **Aerial View** window

The Pan tool

The **Pan** tools can be called with a *click* on the **Pan** button in the status bar, from the **Pan** sub-menu of the **View** drop-down menu or by *entering* **p** at the command line. When the tool is called, the cursor on screen changes to an icon of a hand. *Dragging* the hand across screen under mouse movement allows various parts of a large drawing not in the AutoCAD drawing area to be viewed. As the *dragging* takes place, the green rectangle in the **Aerial View** window moves in sympathy (see Fig. 4.7). The **Pan** tool allows any part of the drawing to be viewed and/or modified. When that part of the drawing which is required is on screen a *right-click* calls up the menu as shown in Fig. 4.7, from which either the tool can be exited or other tools can be called.

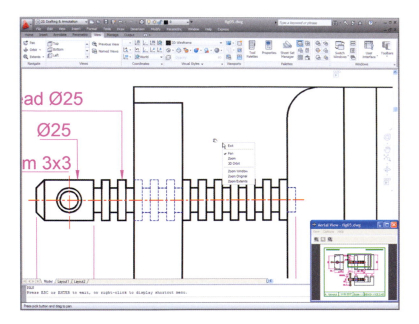

Fig. 4.7 The **Pan** tool in action showing a part of the drawing, while the whole drawing is shown in the **Aerial View** window.

Notes

1. If using a mouse with a wheel both zooms and pans can be performed with the aid of the wheel. See page 8.

2. The **Zoom** tools are important in that they allow even the smallest parts of drawings to be examined and, if necessary, amended or modified.

3. The zoom tools can be called from the sub-menu of the **View** drop-down menu or by *entering* **zoom** or **z** at the command line. The easiest of this choice is to *enter* **z** at the command line followed by a *right-click*.

4. Similarly the easiest method of calling the **Pan** tool is to *enter* **p** at the command line followed by a *right-click*.

5. When constructing large drawings, the **Pan** tool and the **Aerial View** window are of value for allowing work to be carried out in any part of a drawing, while showing the whole drawing in the **Aerial View** window.

Drawing templates

In Chapters 1–3, drawings were constructed in the template **acadiso.dwt** which loaded when AutoCAD 2011 was opened. The default **acadiso** template has been amended to **Limits** set to **420,297** (coordinates within which a drawing can be constructed), **Grid Display** set to **10**, **Snap Mode** set to **5** and the drawing area **Zoomed** to **All**.

Throughout this book most drawings will be based on an **A3** sheet, which measures 420 units by 297 units (the same as **Limits**).

> **Note**
>
> As mentioned before if others are using the computer on which drawings are being constructed, it is as well to save the template being used to another file name, or if thought necessary to a memory stick or other temporary type of disk. A file name **My_template.dwt**, as suggested earlier, or a name such as **book_template** can be given.

Adding features to the template

Four other features will now be added to our template:

Text style – set in the **Text Style** dialog.
Dimension style – set in the **Dimension Style Manager** dialog.
Shortcutmenu variable – set to **0**.
Layers – set in the **Layer Properties Manager** dialog.

Setting text

1. At the command line:

Command: *enter* st (Style) *right-click*

2. The **Text style** dialog appears (Fig. 4.8). In the dialog, *enter* **6** in the **Height** field. Then *left-click* on **Arial** in the **Font name** popup list. **Arial** font letters appear in the **Preview** area of the dialog.

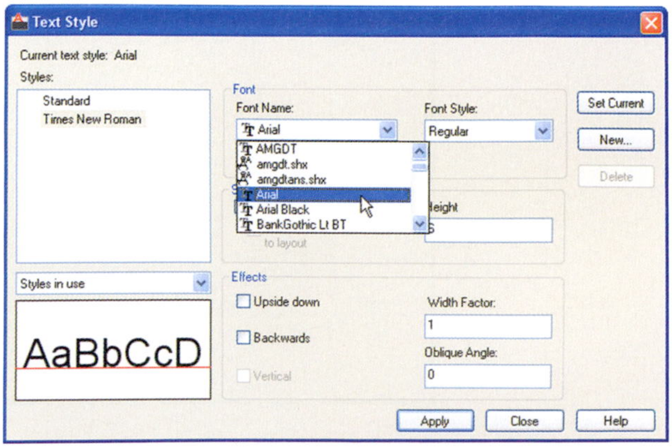

Fig. 4.8 The **Text Style** dialog

3. *Left-click* the **New** button and *enter* **Arial** in the **New** text style sub-dialog which appears (Fig. 4.9) and *click* the **OK** button.
4. *Left-click* the **Set Current** button of the **Text Style** dialog.
5. *Left-click* the **Close** button of the dialog.

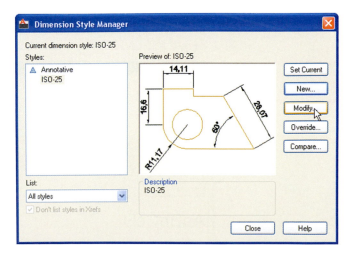

Fig. 4.9 The **New Text Style** sub-dialog

Setting dimension style

Settings for dimensions require making *entries* in a number of sub-dialogs in the **Dimension Style Manager**. To set the dimensions style:

1. At the command line:

Command: *enter* d *right-click*

And the **Dimensions Style Manager** dialog appears (Fig. 4.10).

Fig. 4.10 The **Dimensions Style Manager** dialog

2. In the dialog, *click* the **Modify…** button.
3. The **Modify Dimension Style** dialog appears (Fig. 4.11). This dialog shows a number of tabs at the top of the dialog. *Click* the **Lines** tab and make settings as shown in Fig. 4.11. Then *click* the **OK** button of that dialog.

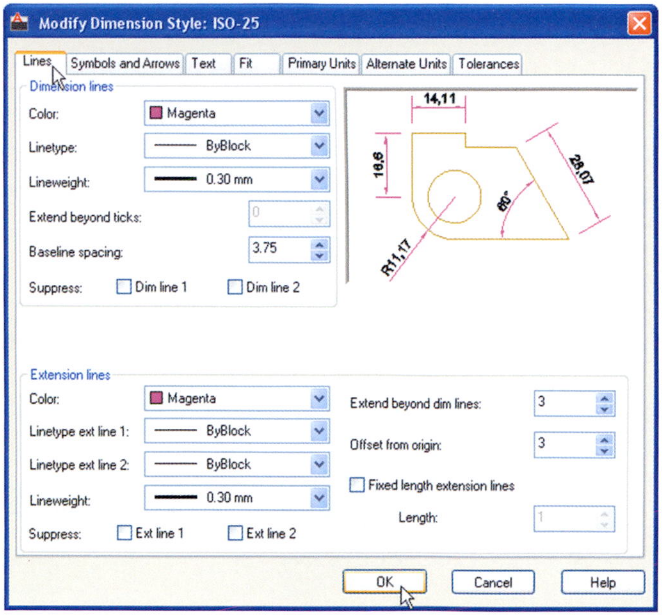

Fig. 4.11 The setting for Lines in the Modify Dimensions Style dialog

4. The original **Dimension Style Manager** reappears. *Click* its **Modify** button again.
5. The **Modify Dimension Style** dialog reappears (Fig. 4.12), *click* the **Symbols and Arrows** tab. Set **Arrow** size to **6**.
6. Then *click* the **Text** tab. Set **Text style** to **Arial**, set **Color** to **Magenta**, set **Text Height** to **6** and *click* the **ISO** check box in the bottom right-hand corner of the dialog.
7. Then *click* the **Primary Units** tab and set the units **Precision** to **0**, that is no units after decimal point and **Decimal separator** to **Period**. *Click* the sub-dialogs **OK** button (Fig. 4.12).
8. The **Dimension Styles Manager** dialog reappears showing dimensions, as they will appear in a drawing, in the **Preview of my-style** box. *Click* the **New…** button. The **Create New Dimension Style** dialog appears (Fig. 4.13).
9. *Enter* a suitable name in the **New Style Name:** field – in this example this is **My-style**. *Click* the **Continue** button and the **Dimension Style Manager** appears (Fig. 4.14). This dialog now shows a preview of the **My-style** dimensions. *Click* the dialog's **Set Current** button, following by another *click* on the **Close** button. See Fig. 4.14.

CHAPTER 4

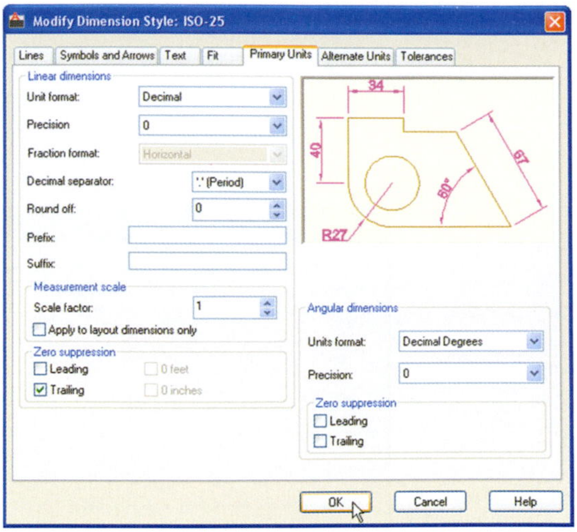

CHAPTER 4

Fig. 4.12 Setting **Primary Units** in the **Dimension Style Manager**

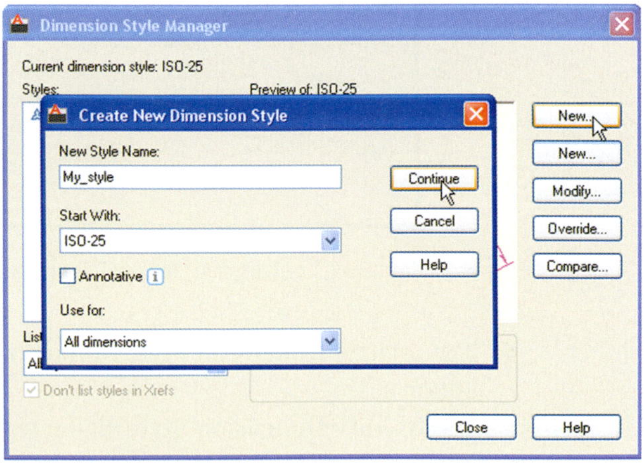

Fig. 4.13 The **Create New Dimension Style** dialog

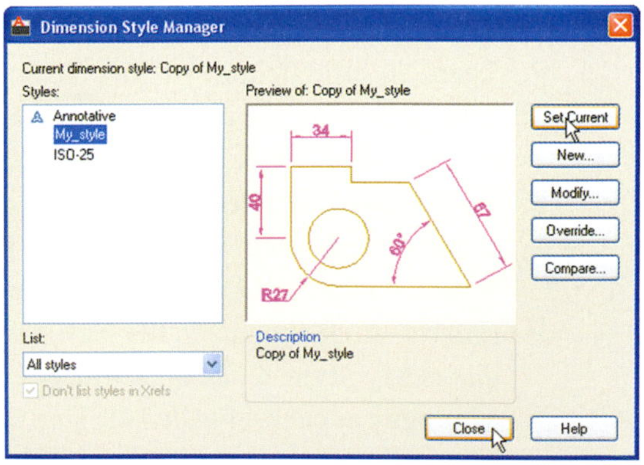

Fig. 4.14 The **Dimension Style Manager** reappears. *Click* the **Set Current** and **Close** buttons

Setting the shortcutmenu variable

Call the line tool, draw a few lines and then *right-click*. The *right-click* menu shown in Fig. 4.15 may well appear. A similar menu will also appear when any tool is called. Some operators prefer using this menu when constructing drawings. To stop this menu appearing:

```
Command: enter shortcutmenu right-click
Enter new value for SHORTCUTMENU <12>: 0
Command:
```

And the menu will no longer appears when a tool is in action.

Fig. 4.15 The right-click menu

Setting layers

1. At the command line *enter* **layer** or **la** followed by a *right-click*. The **Layer Properties Manager** palette appears (Fig. 4.16).
2. *Click* the **New Layer** icon. **Layer1** appears in the layer list. Overwrite the name **Layer1** *entering* **Centre**.

Fig. 4.16 The **Layer Properties Manager** palette

3. Repeat step **2** four times and make four more layers entitled **Construction**, **Dimensions**, **Hidden** and **Text**.
4. *Click* one of the squares under the **Color** column of the dialog. The **Select Color** dialog appears (Fig. 4.17). *Double-click* on one of the colours in the **Index Color** squares. The selected colour appears against the layer name in which the square was selected. Repeat until all five new layers have a colour.

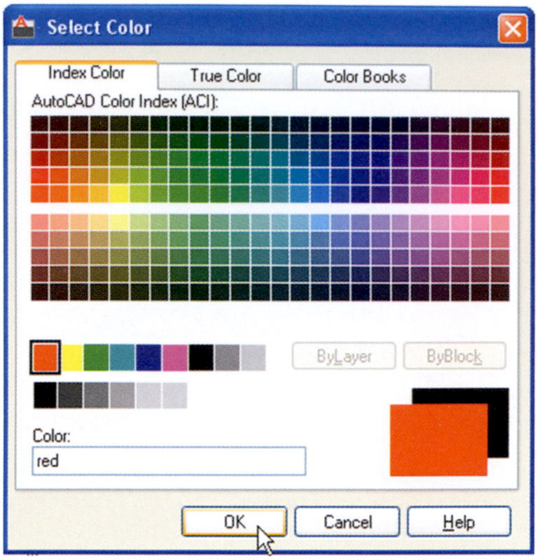

Fig. 4.17 The **Select Color** dialog

5. *Click* on the linetype **Continuous** against the layer name **Centre**.
   The **Select Linetype** dialog appears (Fig. 4.18). *Click* its **Load…**
   button and from the **Load or Reload Linetypes** dialog *double-click*
   CENTER2. The dialog disappears and the name appears in the **Select
   Linetype** dialog. *Click* the **OK** button and the linetype **CENTER2**
   appears against the layer **Center**.

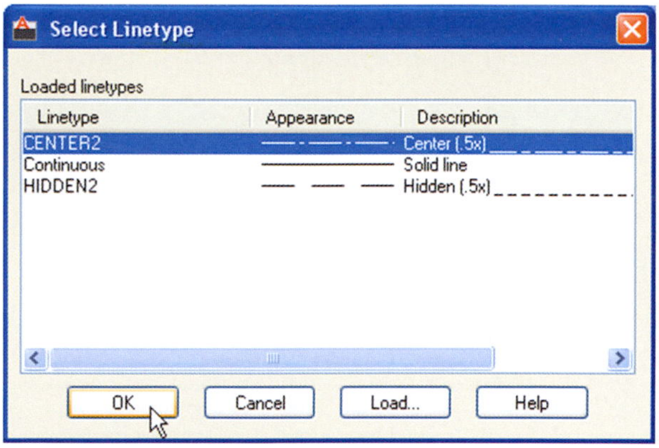

Fig. 4.18 The **Select Linetype** dialog

6. Repeat with layer **Hidden**, load the linetype **HIDDEN2** and make the
   linetype against this layer **HIDDEN2**.
7. *Click* on the any of the lineweights in the **Layer Properties Manager**.
   This brings up the **Lineweight** dialog (Fig. 4.19). Select the lineweight
   0.3. Repeat the same for all other layers. Then *click* the **Close** button of
   the **Layer Properties Manager**.

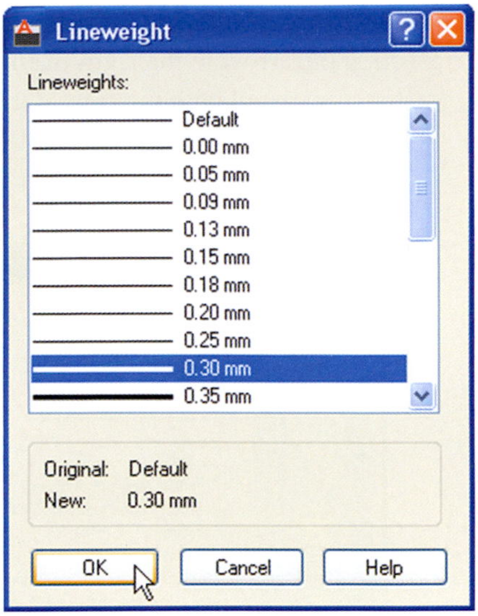

Fig. 4.19 The **Lineweight** dialog

Saving the template file

1. *Left-click* on **Save As** in the menu appearing with a *left-click* on the AutoCAD icon at the top left-hand corner of the screen (Fig. 4.20).

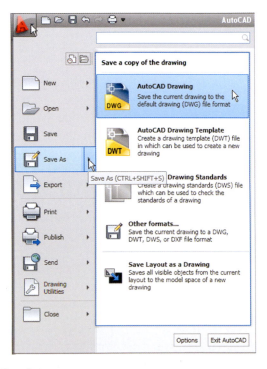

Fig. 4.20 Calling **Save As**

2. In the **Save Drawing As** dialog which comes on screen (Fig. 4.21), *click* the arrow to the right of the **Files of type** field and in the popup list associated with the field *click* on **AutoCAD Drawing Template (*.dwt)**. The list of template files in the **AutoCAD 2011/Template** directory appears in the file list.

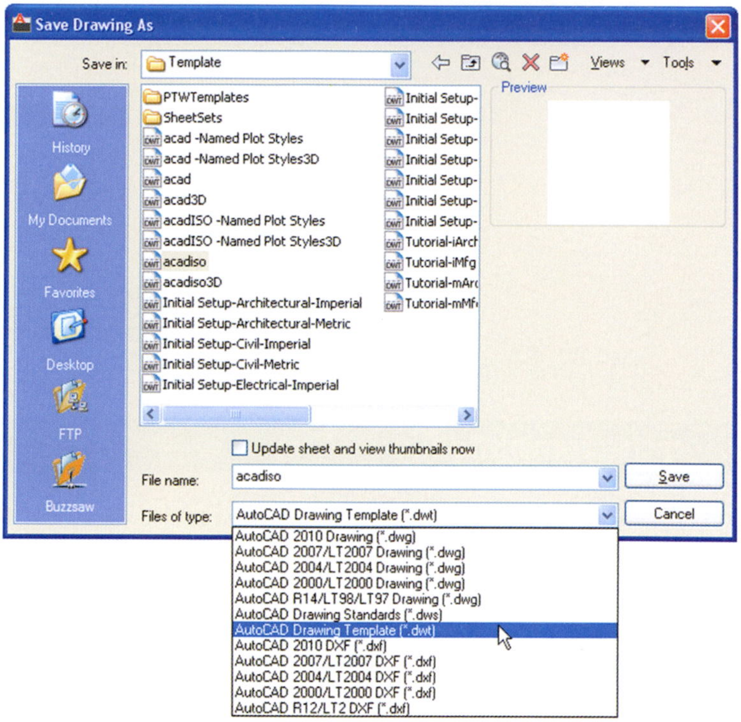

Fig. 4.21 Saving the template to the name **acadiso.dwt**

3. *Click* on **acadiso** in the file list, followed by a *click* on the **Save** button.
4. The **Template Option** dialog appears. Make *entries* as suggested in Fig. 4.22, making sure that **Metric** is chosen from the popup list. The

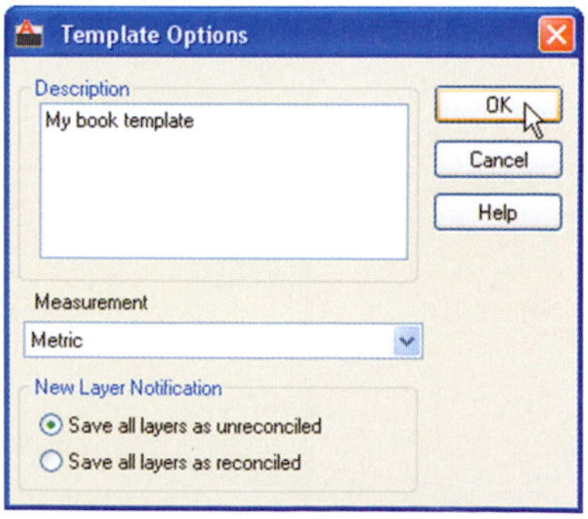

Fig. 4.22 The **Template Description** dialog

template can now saved to be opened for the construction of drawings as needed. Now when AutoCAD 2011 is opened again the template **acadiso.dwt** appears on screen.

> **Note**
>
> Remember that if others are using the computer it is advisable to save the template to a name of your own choice.

Template file to include Imperial dimensions

If dimensions are to be in **Imperial** measure – in yards, feet and inches, first set **Limits** to **28,18**. In addition the settings in the **Dimension Style Manager** will need to be different from those shown earlier. Settings for **Imperial measure** in the **Primary Units** sub-dialog need to be set as shown in Fig. 4.23. Settings in the **Text** sub-dialog of the **Text Style** dialog also need to be set as shown in Fig. 4.24.

In addition the settings in the **Primary Units** dialog also need settings to be different to those for Metric dimensions as shown in Fig. 4.25.

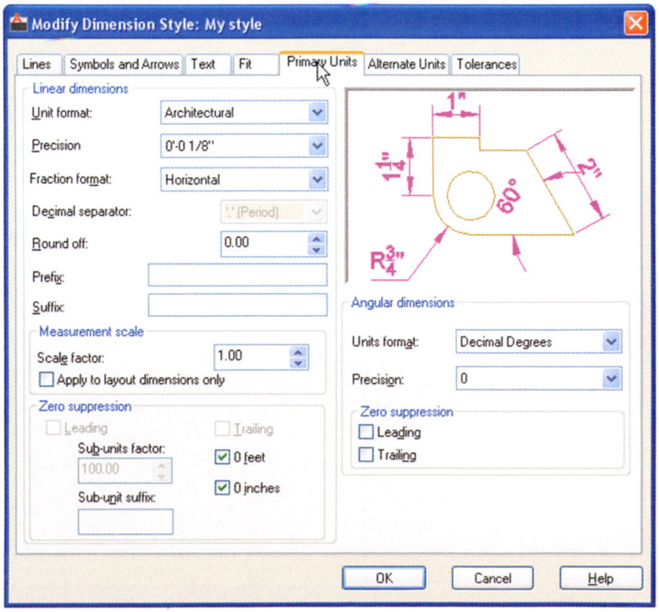

Fig. 4.23 Settings for Imperial dimensions in **Primary Units**

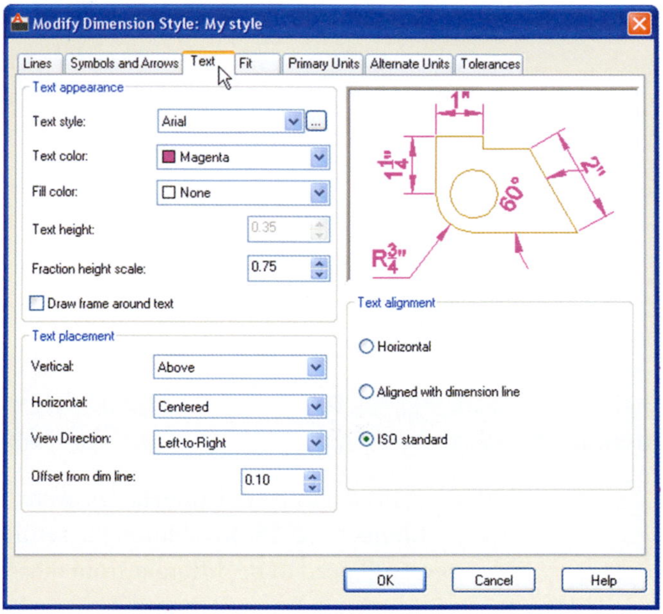

Fig. 4.24 Settings for Imperial dimensions set in **Text**

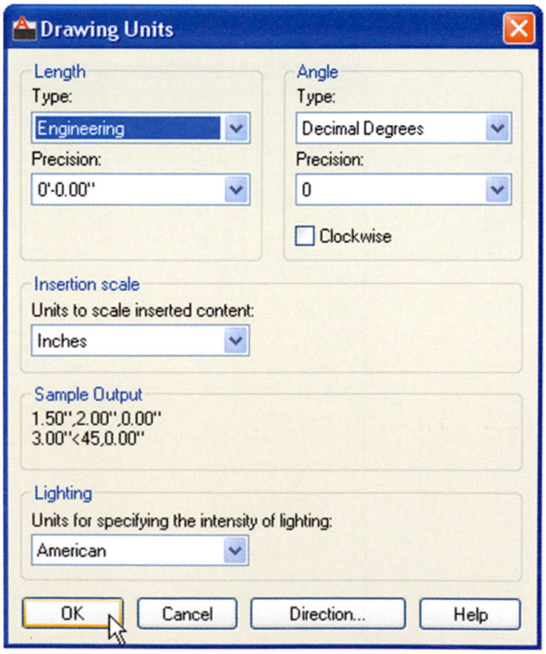

Fig. 4.25 Settings for Imperial dimensions in the **Primary Units** dialog

REVISION NOTES

1. The Zoom tools are important in that they allow even the smallest parts of drawings to be examined, amended or modified.
2. The Zoom tools can be called from the sub-menu of the View drop-down menu, or by entering z or zoom at the command line. The easiest is to enter z at the command line.
3. There are five methods of calling tools for use – selecting a tool icon in a panel from a group of panels in the Ribbon; entering the name of a tool in full at the command line; entering an abbreviation for a tool; selecting a tool from a drop-down menu. If working in the AutoCAD Classic workspace, tools are called from toolbars.
4. When constructing large drawings, the Pan tool and the Aerial View window allow work to be carried out in any part of a drawing, while showing the whole drawing in the Aerial View window.
5. An A3 sheet of paper is 420 mm by 297 mm. If a drawing constructed in the template acadiso.dwt described in this book, is printed/plotted full size (scale 1:1), each unit in the drawing will be 1 mm in the print/plot.
6. When limits are set it is essential to call Zoom followed by a (All) to ensure that the limits of the drawing area are as set.
7. If the right-click menu appears when using tools, the menu can be aborted if required by setting the SHORTCUTMENU variable to 0.

CHAPTER 4

Exercises

If you have saved drawings constructed either by following the worked examples in this book or by answering exercises in Chapters 2 and 3, open some of them and practise zooms and pans.

The Modify tools

AIM OF THIS CHAPTER

The aim of this chapter is to describe the uses of tools for modifying parts of drawings.

Introduction

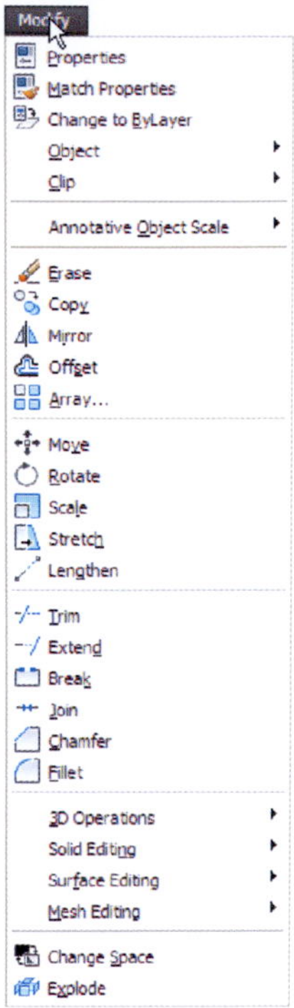

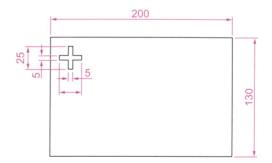

The **Modify** tools are among those most frequently used. The tools are found in the **Home/Modify** panel. A *click* on the arrow at the bottom of the **Home/Modify** panel brings down a further set of tool icons (Fig. 5.1). They can also be selected from the **Modify** drop-down menu (Fig. 5.2). In the **AutoCAD Classic** workspace, they can be selected from the **Modify** toolbar.

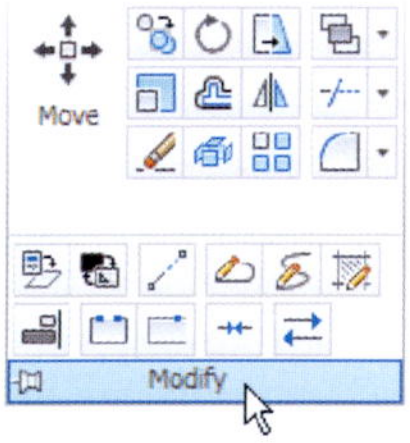

Fig. 5.1 The **Modify** tool icons in the **Home/Modify** panel

Fig. 5.2 The **Modify** drop-down menu

Using the **Erase** tool from **Home/Modify** was described in Chapter 2. Examples of tools other than the **Explode** follow. See also Chapter 9 for **Explode**.

First example – Copy (Fig. 5.5)

1. Construct Fig. 5.3 using **Polyline**. Do not include the dimensions.

Fig. 5.3 First example – **Copy Object** – outlines

The Copy tool

2. Call the **Copy** tool – either *left-click* on its tool icon in the **Home/ Modify** panel (Fig. 5.4) or *enter* **cp** or **copy** at the command line. The command line shows:

```
Command: _copy
Select objects: pick the cross 1 found
Select objects: right-click
```

```
Current settings: Copy mode = Multiple
Specify base point or [Displacement/mOde]
  <Displacement>: pick
Specify second point or [Exit/Undo]: pick
Specify second point or [Exit/Undo] <Exit>:
  right-click
Command:
```

The result is given in Fig. 5.5.

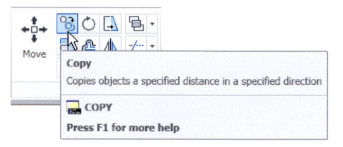

Fig. 5.4 The **Copy** tool from the **Home/Modify** panel

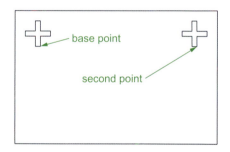

Fig. 5.5 First example – **Copy**

Second example – Multiple copy (Fig. 5.6)

1. Erase the copied object.
2. Call the **Copy** tool. The command line shows:

```
Command: _copy
Select objects: pick the cross 1 found
Select objects: right-click
Current settings: Copy mode = Multiple
Specify base point or [Displacement/mOde]
  <Displacement>: pick
Specify second point or <use first point as
  displacement>: pick
```

```
Specify second point or [Exit/Undo] <Exit>: pick
Specify second point or [Exit/Undo] <Exit>: pick
Specify second point or [Exit/Undo] <Exit>: e
  (Exit)
Command
```

The result is shown in Fig. 5.6.

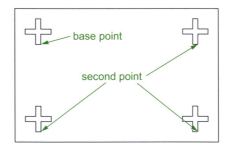

Fig. 5.6 Second example – **Copy** – **Multiple copy**

The Mirror tool

First example – Mirror (Fig. 5.9)

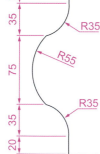

1. Construct the outline Fig. 5.7 using the **Line** and **Arc** tools.
2. Call the **Mirror** tool – *left-click* on its tool icon in the **Home/Modify** panel (Fig. 5.8) or from the **Modify** drop-down menu, or *enter* **mi** or **mirror** at the command line. The command line shows:

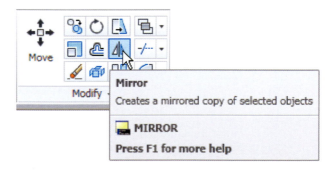

Fig. 5.7 First example – **Mirror** – outline

Fig. 5.8 The **Mirror** tool from the **Home/Modify** panel

```
Command:_mirror
Select objects: pick first corner Specify opposite
  corner: pick 7 found
Select objects: right-click
Specify first point of mirror line: pick
```

```
Specify second point of mirror line: pick
Erase source objects [Yes/No] <N>: right-click
Command:
```

The result is shown in Fig. 5.9.

Second example – Mirror (Fig. 5.10)

1. Construct the outline shown in the dimensioned polyline in the upper drawing of Fig. 5.10.
2. Call **Mirror** and using the tool three times complete the given outline. The two points shown in Fig. 5.10 are to mirror the right-hand side of the outline.

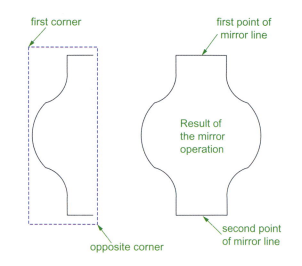

Fig. 5.9 First example – **Mirror**

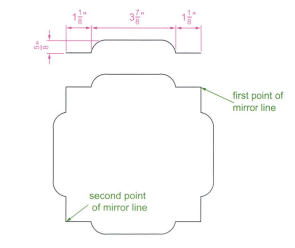

Fig. 5.10 Second example – **Mirror**

Third example – Mirror (Fig. 5.11)

Fig. 5.11 Third example – **Mirror**

If text is involved when using the **Mirror** tool, the set variable **MIRRTEXT** must be set correctly. To set the variable:

```
Command: mirrtext
Enter new value for MIRRTEXT <1>: 0
Command:
```

If set to **0** text will mirror without distortion. If set to **1** text will read backwards as indicated in Fig. 5.11.

The Offset tool

Examples – Offset (Fig. 5.14)

1. Construct the four outlines shown in Fig. 5.13.
2. Call the **Offset** tool – *left-click* on its tool icon in the **Home/Modify** panel (Fig. 5.12), *pick* the tool name in the **Modify** drop-down menu or *enter* **o** or **offset** at the command line. The command line shows (Fig. 5.13):

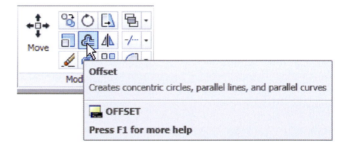

Fig. 5.12 The **Offset** tool from the **Home/Modify** panel

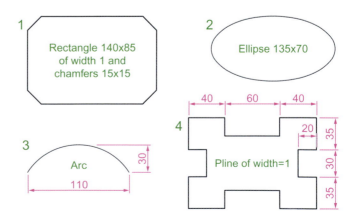

Fig. 5.13 Examples – **Offset** – outlines

```
Command:_offset
Current settings: Erase source = No
  Layer=Source OFFSETGAPTYPE=0
Specify offset distance or [Through/Erase/Layer]
  <Through>: 10
Select object to offset or [Exit/Undo]
  <Exit>: pick drawing 1
Specify point on side to offset or [Exit/Multiple/
  Undo] <Exit>: pick inside the rectangle
Select object to offset or [Exit/Undo]
  <Exit>: e (Exit)
Command:
```

3. Repeat for drawings **2**, **3** and **4** in Fig. 5.12 as shown in Fig. 5.14.

Arrays can be in either a **Rectangular** form or a **Polar** form as shown in the examples below.

CHAPTER 5

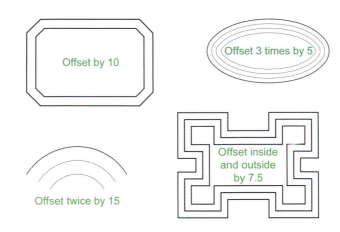

Offset by 10

Offset 3 times by 5

Offset inside and outside by 7.5

Offset twice by 15

Fig. 5.14 Examples – **Offset**

The Array tool

First example – Rectangular Array (Fig. 5.17)

1. Construct the drawing Fig. 5.15.
2. Call the **Array** tool – either *click* **Array** in the **Modify** drop-down menu (Fig. 5.16), from the **Home/Modify** panel, or *enter* **ar** or **array** at the command line. The **Array** dialog appears (Fig. 5.17).
3. Make settings in the dialog:
   Rectangular Array radio button set on (dot in button)
   Row field – *enter* **5**

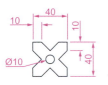

Fig. 5.15 First example – **Array** – drawing to be arrayed

Column field – *enter* **6**
Row offset field – *enter* – **50** (note the minus sign)
Column offset field – *enter* **50**

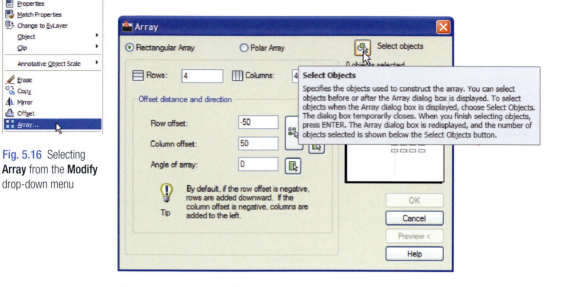

Fig. 5.16 Selecting
Array from the **Modify**
drop-down menu

Fig. 5.17 First example – the **Array** dialog

4. *Click* the **Select objects** button and the dialog disappears. Window the drawing. A second dialog appears which includes a **Preview**< button.
5. *Click* the **Preview**< button. The dialog disappears and the following prompt appears at the command line:

```
Pick or press Esc to return to drawing or <Right-
    click to accept drawing>:
```

6. If satisfied *right-click*. If not, press the **Esc** key and make revisions to the **Array** dialog fields as necessary.

The resulting array is shown in Fig. 5.18.

Second example – Polar Array (Fig. 5.22)

1. Construct the drawing Fig. 5.19.
2. Call **Array**. The **Array** dialog appears. Make settings as shown in Fig. 5.20.
3. *Click* the **Select objects** button of the dialog and window the drawing. The dialog returns to screen. *Click* the **Pick Center point** button (Fig. 5.21) and when the dialog disappears, *pick* a centre point for the array.

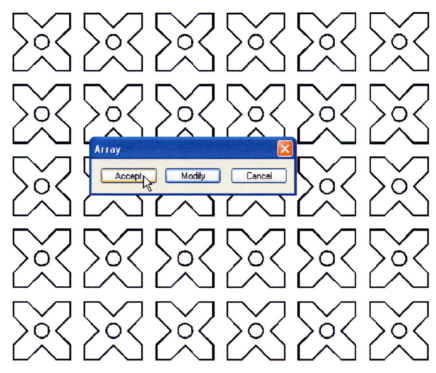

Fig. 5.18 First example – **Array**

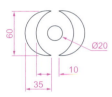

Fig. 5.19 Second
example – the drawing
to be arrayed

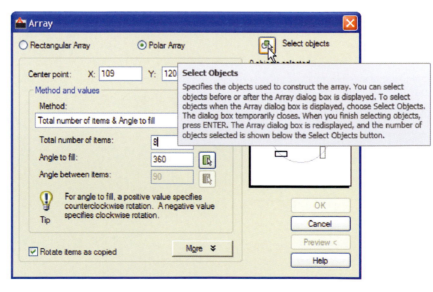

Fig. 5.20 Second example – **Array** – settings in the dialog

Array dialog box

○ Rectangular Array   ● Polar Array   [⊡] Select objects

1 objects selected

Center point: X: 202   Y: 263   [⊡]

Method and values

Method:

Total number of items & Angle to fill

Total number of items: 8

Angle to fill: 360

Angle between items: 90

Select Objects

Specifies the objects used to construct the array. You can select objects before or after the Array dialog box is displayed. To select objects when the Array dialog box is displayed, choose Select Objects. The dialog box temporarily closes. When you finish selecting objects, press ENTER. The Array dialog box is redisplayed, and the number of objects selected is shown below the Select Objects button.

💡 For angle to fill, a positive value specifies counterclockwise rotation. A negative value specifies clockwise rotation.
Tip

☑ Rotate items as copied   More ❯

[OK]
[Cancel]
[Preview <]
[Help]

Fig. 5.21 Second example – **Array** – the **Pick Center point** button

4. The dialog reappears. *Click* its **Preview**< button. The array appears and the command line shows:

```
Pick or press Esc to return to drawing or
  <Right-click to accept drawing>:
```

5. If satisfied *right-click*. If not, press the **Esc** key and make revisions to the **Array** dialog fields as necessary.

The resulting array is shown in Fig. 5.22.

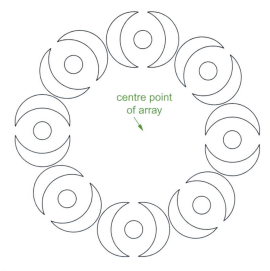

centre point
of array

Fig. 5.22 Second example – **Array**

The Move tool

Example – Move (Fig. 5.25)

1. Construct the drawing Fig. 5.23.

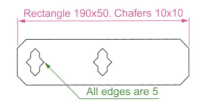

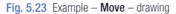

Rectangle 190x50. Chafers 10x10

All edges are 5

Fig. 5.23 Example – **Move** – drawing

2. Call **Move** – *click* the **Move** tool icon in the **Home/Modify** panel (Fig. 5.24), *pick* **Move** from the **Modify** drop-down menu or *enter* **m** or **move** at the command line, which shows:

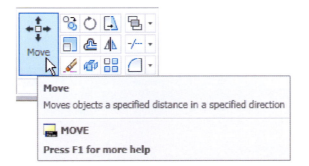

Fig. 5.24 The **Move** tool from the **Home/Modify** panel

```
Command:_move
Select objects: pick the middle shape in the
  drawing 1 found
Select objects: right-click
Specify base point or [Displacement]
  <Displacement>: pick
Specify second point or <use first point as
  displacement>: pick
Command:
```

The result is given in Fig. 5.25.

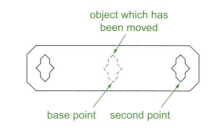

object which has
been moved

base point   second point

Fig. 5.25 Example – **Move**

The Rotate tool

When using the **Rotate** tool remember the default rotation of objects within AutoCAD 2011 is counterclockwise (anticlockwise).

Example – Rotate (Fig. 5.27)

1. Construct drawing **1** of Fig. 5.27 with **Polyline**. Copy the drawing **1** three times (Fig. 5.27).
2. Call **Rotate** – *left-click* on its tool icon in the **Home/Modify** panel (Fig. 5.26), *pick* **Rotate** from the **Modify** drop-down menu or *enter* **ro** or **rotate** at the command line. The command line shows: and the first copy rotates through the specified angle.

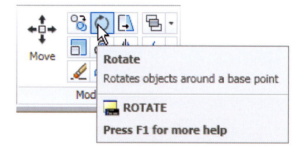

Fig. 5.26 The **Rotate** tool icon from the **Home/Modify** panel

```
Command:_rotate
Current positive angle in UCS:
  ANGDIR = counterclockwise ANGBASE=0
Select objects: window the drawing 3 found
Select objects: right-click
Specify base point: pick
Specify rotation angle or [Copy/Reference] <0>: 45
Command:
```

3. Repeat for drawings **3** and **4** rotating as shown in Fig. 5.27.

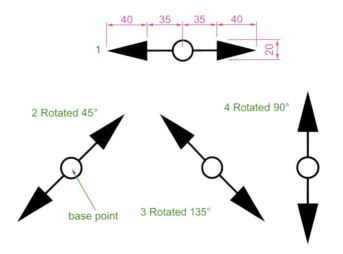

Fig. 5.27 Example – **Rotate**

The Scale tool

Examples – **Scale (Fig. 5.29)**

1. Using the **Rectangle and** Polyline tools, construct drawing **1** of
    Fig. 5.29. The **Rectangle** fillets are R10. The line width of all parts
    is **1**. Copy the drawing 3 times to give drawings **2**, **3** and **4**.
2. Call **Scale** – *left-click* on its tool icon in the **Home/Draw** panel
    (Fig. 5.28), *pick* **Scale** from the **Modify** drop-down-menu or *enter* **sc** or
    scale at the command line which then shows:

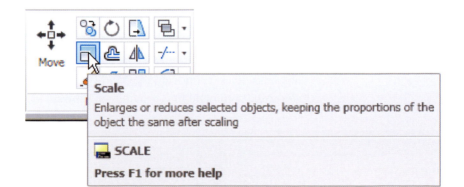

Fig. 5.28 The **Scale** tool from the **Home/Modify** panel

```
Command:_scale
Select objects: window drawing 2 5 found
Select objects: right-click
Specify base point: pick
```

```
Specify scale factor or [Copy/Reference]
  <1>: 0.75
Command:
```

3. Repeat for the other two drawings **3** and **4** scaling to the scales given with the drawings.

The results are shown in Fig. 5.29.

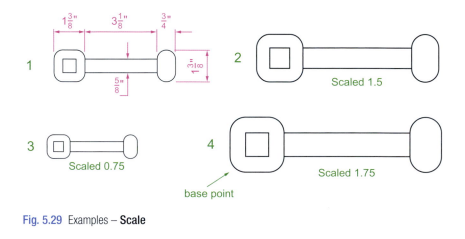

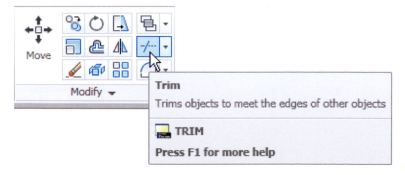

Fig. 5.29 Examples – **Scale**

The Trim tool

This tool is one which will be frequent use when constructing drawings.

First example – Trim (Fig. 5.31)

1. Construct the drawing **Original drawing** in Fig. 5.31.
2. Call **Trim** – either *left-click* on its tool icon in the **Home/Modify** panel (Fig. 5.30), *pick* **Trim** from the **Modify** drop-down menu or *enter* **tr** or **trim** at the command line, which then shows:

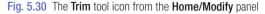

Fig. 5.30 The **Trim** tool icon from the **Home/Modify** panel

```
Command:_trim
Current settings: Projection UCS. Edge = Extend
Select cutting edges ....
Select objects or <select all>: pick the left-hand
  circle 1 found
Select objects: right-click
Select objects to trim or shift-select to extend
  or [Fence/Project/Crossing/Edge/eRase//Undo]:
  pick one of the objects
Select objects to trim or shift-select to extend
  or
[Fence/Crossing/Project/Edge/eRase/Undo: pick the
  second of the objects
Select objects to trim or shift-select to extend
  or [Project/Edge/Undo]: right-click
Command:
```

3. This completes the **First stage** as shown in Fig. 5.31. Repeat the **Trim** sequence for the **Second stage**.
4. The **Third stage** drawing of Fig. 5.31 shows the result of the trims at the left-hand end of the drawing.
5. Repeat for the right-hand end. The final result is shown in the drawing labelled **Result** in Fig. 5.31.

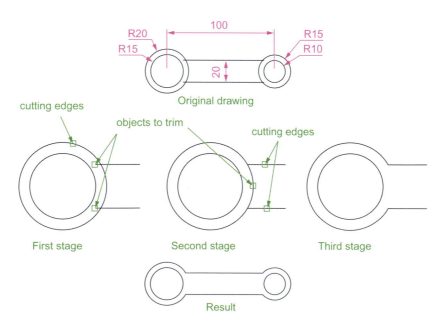

Fig. 5.31 First example – **Trim**

CHAPTER 5

Second example – Trim (Fig. 5.32)

1. Construct the left-hand drawing of Fig. 5.32.
2. Call **Trim**. The command line shows:

```
Command:_trim
Current settings: Projection UCS. Edge = Extend
Select cutting edges ...
Select objects or <select all>: pick the left-hand
  arc 1 found
Select objects: right-click
Select objects to trim or shift-select to extend
  or [Fence/Crossing/Project/Edge/eRase/Undo]: e
  (Edge)
Enter an implied edge extension mode [Extend/No
  extend] <No extend>: e (Extend)
Select objects to trim: pick
Select objects to trim: pick
Select objects to trim: right-click
Command:
```

3. Repeat for the other required trims. The result is given in Fig. 5.32.

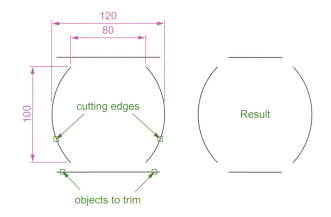

Fig. 5.32 Second example – **Trim**

The Stretch tool

Examples – Stretch (Fig. 5.34)

As its name implies the **Stretch** tool is for stretching drawings or parts of drawings. The action of the tool prevents it from altering the shape of circles in any way. Only **crossing** or **polygonal** windows can be used to determine the part of a drawing which is to be stretched.

1. Construct the drawing labelled **Original** in Fig. 5.34, but do not include the dimensions. Use the **Circle**, **Arc**, **Trim** and **Polyline Edit** tools. The resulting outlines are plines of width = 1. With the **Copy** tool make two copies of the drawing.

Note

In each of the three examples in Fig. 5.34, the broken lines represent the crossing windows required when **Stretch** is used.

---

2. Call the **Stretch** tool – either *click* on its tool icon in the **Home/Modify** panel (Fig. 5.33), *pick* its name in the **Modify** drop-down menu or *enter* **s** or **stretch** at the command line, which shows.

Fig. 5.33 The **Stretch** tool icon from the **Home/Modify** panel

```
Command:_stretch
Select objects to stretch by crossing-window
  or crossing-polygon...
Select objects:enter c right-click
Specify first corner: pick Specify opposite
  corner: pick 1 found
Select objects: right-click
Specify base point or [Displacement]
  <Displacement>: pick beginning of arrow
Specify second point of displacement or <use first
  point as displacement>: drag in the direction
  of the arrow to the required second point and
  right-click
Command:
```

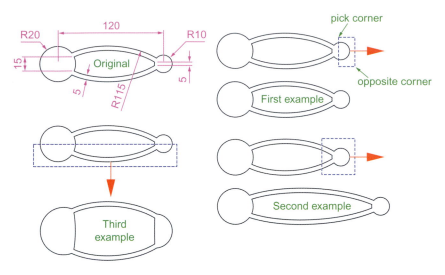

Fig. 5.34 Examples – **Stretch**

Notes

1. When circles are windowed with the crossing window no stretching can take place. This is why, in the case of the first example in Fig. 5.33, when the **second point of displacement** was *picked*, there was no result – the outline did not stretch.

2. Care must be taken when using this tool as unwanted stretching can occur (Fig. 5.34).

The Break tool

Examples – Break (Fig. 5.36)

1. Construct the rectangle, arc and circle (Fig. 5.36).
2. Call **Break** – either *click* on its tool icon in the **Home/Modify** panel (Fig. 5.35), *click* **Break** in the **Modify** drop-down menu or *enter* **br** or **break** at the command line, which shows:

For drawings 1 and 2

```
Command:_break Select object: pick at the point
Specify second break point or [First point]: pick
Command:
```

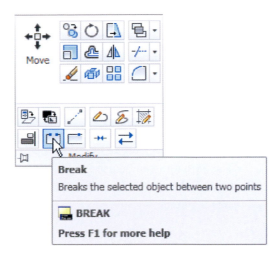

Fig. 5.35 The **Break** tool icon from the **Home/Modify** panel

For drawing 3

```
Command:_break Select object pick
Specify second break point or [First point]: enter
  f right-click
Specify first break point: pick
Specify second break point: pick
Command:
```

The results are shown in Fig. 5.36.

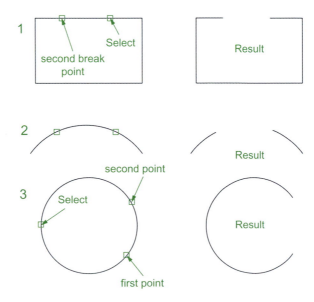

Fig. 5.36 Examples – **Break**

> **Note**
>
> Remember the default rotation of AutoCAD 2011 is counterclockwise. This applies to the use of the **Break** tool.

The Join tool

The **Join** tool can be used to join plines providing their ends are touching, to join lines which are in line with each other, and to join arcs and convert arcs to circles.

Examples – Join (Fig. 5.38)

1. Construct a rectangle from four separate plines – drawing **1** of Fig. 5.38; construct two lines – drawing **2** of Fig. 5.38 and an arc – drawing **3** of Fig. 5.38.
2. Call the **Join** tool – either *click* the **Join** tool icon in the **Home/Modify** panel (Fig. 5.37), select **Join** from the **Modify** drop-down menu or *enter* **join** or **j** at the command line. The command line shows:

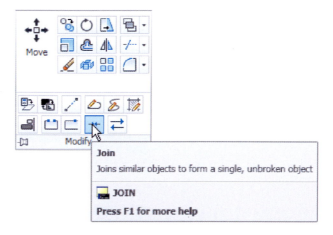

Fig. 5.37 The **Join** tool icon from the **Home/Modify** panel

```
Command: _join Select source object:
Select objects to join to source: pick a pline 1
  found
Select objects to join to source: pick another 1
  found, 2 total
Select objects to join to source: pick another 1
  found, 3 total
Select objects to join to source: right-click
```

```
3 segments added to polyline
Command: right-click
JOIN Select source object: pick one of the lines
Select lines to join to source: pick the other 1
  found
Select lines to join to source: right-click
1 line joined to source
Command: right-click
JOIN Select source object: pick the arc
Select arcs to join to source or [cLose]: enter l
  right-click
Arc converted to a circle.
Command:
```

The results are shown in Fig. 5.38.

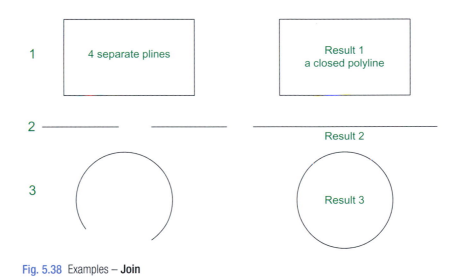

Fig. 5.38 Examples – **Join**

The Extend tool

Examples – **Extend** (Fig. 5.40)

1. Construct plines and a circle as shown in the left-hand drawings of Fig. 5.40.
2. Call **Extend** – either *click* the **Extend** tool icon in the **Home/Modify** panel (Fig. 5.39), *pick* **Extend** from the **Modify** drop-down menu or *enter* **ex** or **extend** at the command line which then shows:

```
Command:_extend
Current settings: Projection=UCS Edge=Extend
Select boundary edges ...
```

CHAPTER 5

```
Select objects or <select all>: pick 1 found
Select objects: right-click
Select object to extend or shift-select to trim
  or[Fence/Crossing/Project/Edge/Undo]: pick
```

Repeat for each object to be extended. Then:

```
Select object to extend or shift-select to trim or
  [Fence/Crossing/Project/Edge/Undo]: right-click
Command:
```

The results are shown in Fig. 5.40.

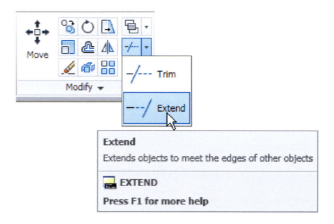

Fig. 5.39 The **Extend** tool icon from the **Home/Modify** panel

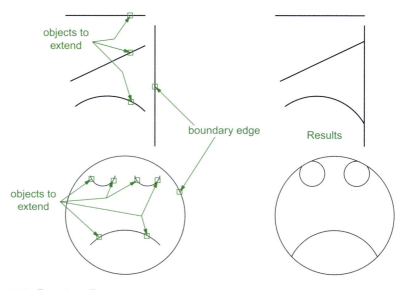

Fig. 5.40 Examples – **Extend**

> **Note**
>
> Observe the similarity of the **Extend** and **No extend** prompts with those of the **Trim** tool.

The Fillet and Chamfer tools

These two tools can be called from the **Home/Modify** panel. There are similarities in the prompt sequences for these two tools. The major differences are that only one (**Radius**) setting is required for a fillet, but two (**Dist1** and **Dist2**) are required for a chamfer. The basic prompts for both are:

Fillet

```
Command:_fillet
Current settings: Mode = TRIM, Radius = 1
Select first object or [Polyline/Radius/Trim/
  mUltiple]: enter r (Radius)right-click
Specify fillet radius <1>: 15
```

Chamfer

```
Command:_chamfer
(TRIM mode) Current chamfer Dist1 = 1, Dist2 = 1
Select first line or [Undo/Polyline/Distance/Angle/
  Trim/mEthod/Multiple]: enter d (Distance)
  right-click
Specify first chamfer distance <1>: 10
Specify second chamfer distance <10>: right-click
```

Examples – Fillet (Fig. 5.42)

1. Construct three rectangles 100 by 60 using either the **Line** or the **Polyline** tool (Fig. 5.42).
2. **Call** Fillet – *click* the arrow to the right of the tool icon in the **Home/Modify** panel and select **Fillet** from the menu which appears (Fig. 5.41), *pick* **Fillet** from the **Modify** drop-down menu or *enter* **f** or **fillet** at the command line which then shows:

```
Command:_fillet
Current settings: Mode = TRIM, Radius = 1
```

```
Select first object or [Polyline/Radius/Trim/
  mUltiple]: r (Radius)
Specify fillet radius <0>: 15
Select first object or [Undo/Polyline/Radius/Trim/
  Multiple]: pick
Select second object or shift-select to apply
  corner: pick
Command:
```

Three examples are given in Fig. 5.42.

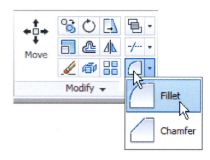

Fig. 5.41 Select **Fillet** from the menu in the **Home/Modify** panel

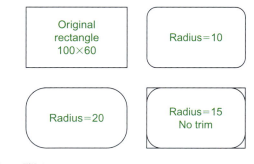

Fig. 5.42 Examples – **Fillet**

Examples – Chamfer (Fig. 5.44)

1. Construct three rectangles 100 by 60 using either the **Line** or the **Polyline** tool.
2. Call **Chamfer** – *click* the arrow to the right of the tool icon in the **Home/Modify** panel and select **Chamfer** from the menu which appears

(Fig. 5.43), *pick* **Chamfer** from the **Modify** drop-down menu or *enter* **cha** or **chamfer** at the command line which then shows:

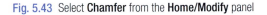

Fig. 5.43 Select **Chamfer** from the **Home/Modify** panel

```
Command:_chamfer
(TRIM mode) Current chamfer Dist1 = 1, Dist2 = 1
Select first line or [Undo/Polyline/Distance/Angle/
  Trim/
mEthod/Multiple]: d
Specify first chamfer distance <1>: 10
Specify second chamfer distance <10>: right-click
Select first line or [Undo/Polyline/Distance/Angle/
  Trim/mEthod/Multiple]:pick the first line for the
  chamfer
Select second line or shift-select to apply
  corner: pick
Command:
```

The result is shown in Fig. 5.44. The other two rectangles are chamfered in a similar manner except that the **No trim** prompt is brought into operation with the bottom left-hand example.

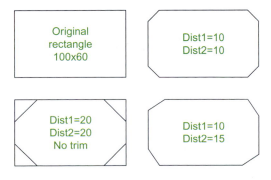

Fig. 5.44 Examples – **Chamfer**

REVISION NOTES

1. The Modify tools are among the most frequently used tools in AutoCAD 2011.
2. The abbreviations for the Modify tools are:
   Copy – cp or co
   Mirror – mi
   Offset – o
   Array – ar
   Move – m
   Rotate – ro
   Scale – sc
   Stretch – s
   Trim – tr
   Extend – ex
   Break – br
   Join – j
   Chamfer – cha
   Fillet – f
3. There are two other tools in the 2D Draw control panel: Erase – some examples were given in Chapter 2 – and Explode – further details of this tools will be given in Chapter 9.

A note – selection windows and crossing windows

In the Options dialog settings can be made in the Selection sub-dialog for Visual Effects. A click on the Visual Effects Settings… button brings up another dialog. If the Area Selection Effect settings are set, on a normal window from top left to bottom right will colour in a chosen colour (default blue). A crossing window from bottom left to top right, will be coloured red. Note also that highlighting – selection Preview Effect allows objects to highlight if this feature is on. These settings are shown in Fig. 5.45.

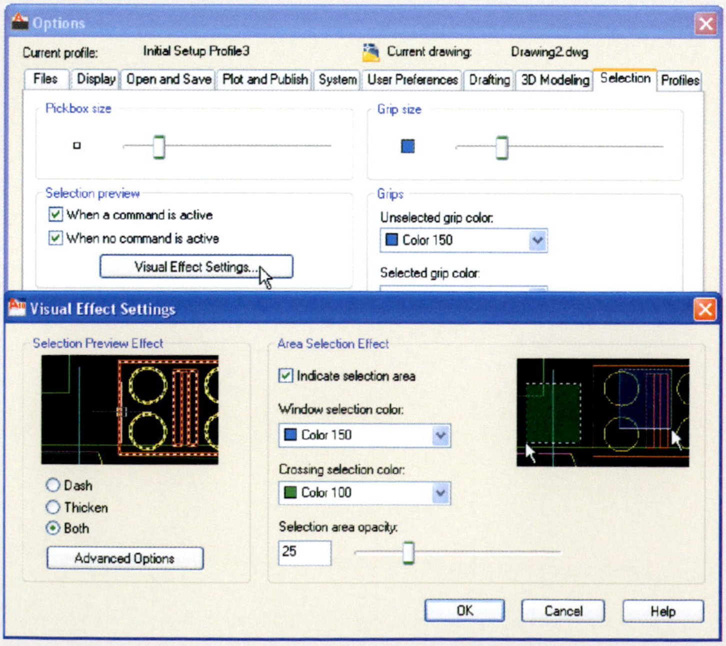

Fig. 5.45 **Visual Setting Effects Settings** sub-dialog of the **Options** dialog

REVISION NOTES (CONTINUED)

4. When using Mirror, if text is part of the area to be mirrored, the set variable Mirrtext will require setting – to either 1 or 0.
5. With Offset the Through prompt can be answered by clicking two points in the drawing area the distance of the desired offset distance.
6. Polar Arrays can be arrays around any angle set in the Angle of array field of the Array dialog.
7. When using Scale, it is advisable to practise the Reference prompt.
8. The Trim tool in either its Trim or its No trim modes is among the most useful tools in AutoCAD 2011.
9. When using Stretch circles are unaffected by the stretching.
10. There are some other tools in the Home/Modify panel not described in this book. The reader is invited to experiment with these other tools. They are:
Bring to Front, Send to Back, Bring above Objects, Send under Objects;
Set by Layer; Change Space; Lengthen; Edit Spline, Edit Hatch; Reverse.

Exercises

Methods of constructing answers to the following exercises can be found in the free website:

http://books.elsevier.com/companions/978-0-08-096575-8

1. Construct the Fig. 5.46. All parts are plines of width = 0.7 with corners filleted R10. The long strips have been constructed using **Circle**, **Polyline**, **Trim** and **Polyline Edit**. Construct one strip and then copy it using **Copy**.

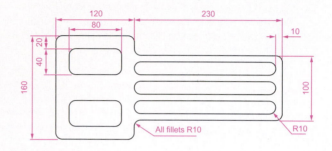

Fig. 5.46 Exercise 1

2. Construct the drawing Fig. 5.47. All parts of the drawing are plines of width = 0.7. The setting in the **Array** dialog is to be **180** in the **Angle of array** field.

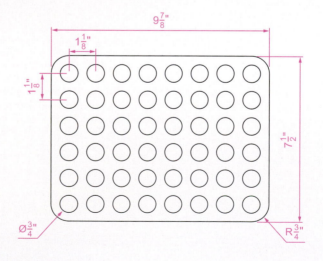

Fig. 5.47 Exercise 2

3. Using the tools **Polyline**, **Circle**, **Trim**, **Polyline Edit**, **Mirror** and **Fillet** construct the drawing (Fig. 5.48).

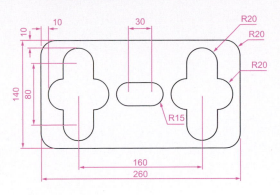

Fig. 5.48 Exercise 3

4. Construct the circles and lines (Fig. 5.49). Using **Offset** and the **Ttr** prompt of the **Circle** tool followed by **Trim**, construct one of the outlines arrayed within the outer circle. Then, with **Polyline Edit** change the lines and arcs into a pline of width = 0.3. Finally array the outline 12 times around the centre of the circles (Fig. 5.50).

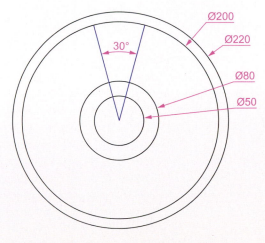

Fig. 5.49 Exercise 4 – circles and lines on which the exercise is based

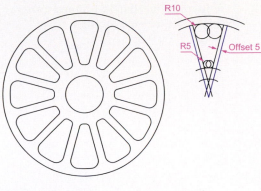

Fig. 5.50 Exercise 4

5. Construct the arrow (Fig. 5.51). Array the arrow around the centre of its circle 8 times to produce the right-hand drawing of Fig. 5.51.

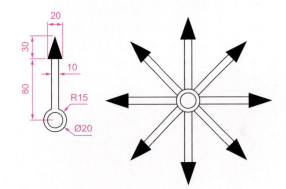

Fig. 5.51 Exercise 5

6. Construct the left-hand drawing of Fig. 5.52. Then with **Move**, move the central outline to the top left-hand corner of the outer outline. Then with **Copy** make copies to the other corners.

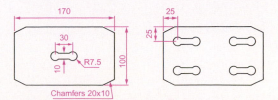

Fig. 5.52 Exercise 6

7.   Construct the drawing Fig. 5.53 and make two copies using **Copy**. With **Rotate** rotate each of the copies to the angles as shown.

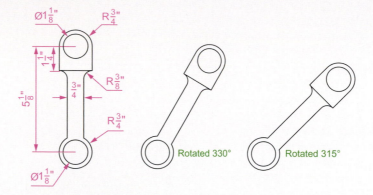

Fig. 5.53 Exercise 7

8.   Construct the dimensioned drawing of Fig. 5.54. With **Copy** copy the drawing. Then with **Scale** scale the drawing to scale of **0.5**, followed by using **Rotate** to rotate the drawing through an angle of as shown. Finally scale the original drawing to a scale of **2:1**.

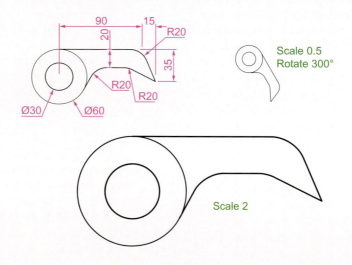

Fig. 5.54 Exercise 8

9. Construct the left-hand drawing of Fig. 5.55. Include the dimensions in your drawing. Then, using the **Stretch** tool, stretch the drawing, including its dimensions to the sizes as shown in the right-hand. The dimensions are said to be **associative** (see Chapter 6).

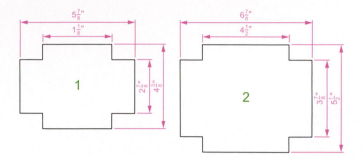

Fig. 5.55 Exercise 9

10. Construct the drawing Fig. 5.56. All parts of the drawing are plines of width = 0.7. The setting in the **Array** dialog is to be **180** in the **Angle of array** field.

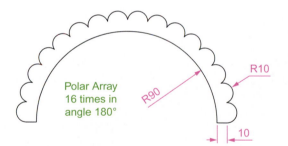

Polar Array
16 times in
angle 180°

Fig. 5.56 Exercise 10

Dimensions and Text

AIMS OF THIS CHAPTER

The aims of this chapter are:

1. To describe a variety of methods of dimensioning drawings.
2. To describe methods of adding text to drawings.

Introduction

The dimension style (**My_style**) has already been set in the acadiso.dwt template, which means that dimensions can be added to drawings using this dimension style.

The Dimension tools

There are several ways in which the dimensions tools can be called.

1. From the **Annotate/Dimensions** panel (Fig. 6.1).

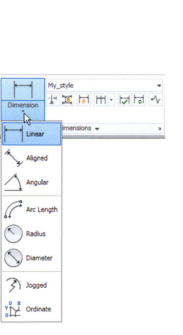

Fig. 6.1 Dimension tools in the **Annotate/Dimensions** panel

Fig. 6.2 Dimensions in the drop-down menu

2. *Click* **Dimension** in the menu bar. Dimension tools can be selected from the drop-down menu which appears (Fig. 6.2).
3. By *entering* an abbreviation for a dimension tool at the command line.

Some operators may well decide to use a combination of the three methods.

4. In the **Classic AutoCAD** workspace from the **Dimension** toolbar.

> **Note**
>
> In general, in this book dimensions are shown in drawings in the Metric style – mainly in millimetres, but some will be shown in Imperial style – in inches. To see how to set a drawing template for Imperial dimensioning see Chapter 4 (page 95).

Adding dimensions using these tools

First example – Linear Dimension (Fig. 6.4)

1. Construct a rectangle 180 × 110 using the **Polyline** tool.
2. Make the **Dimensions** layer current from the **Home/Layers** panel (Fig. 6.3).

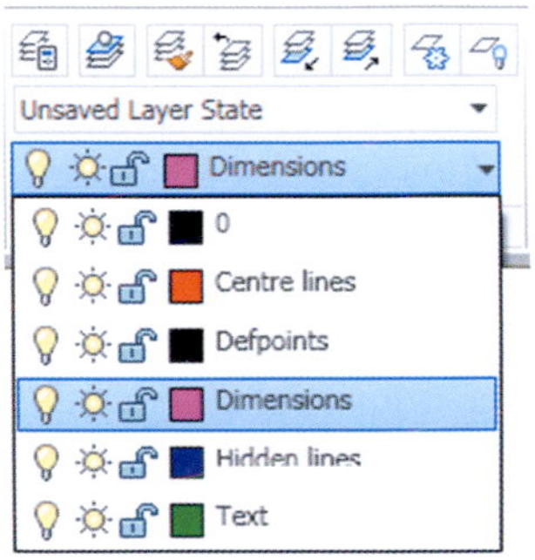

Fig. 6.3 The **Home/Layers** panel – making **Dimensions** layer current

3. *Click* the **Linear** tool icon in the **Annotate/Dimension** panel (Fig. 6.1). The command line shows:

```
Command: _dimlinear
Specify first extension line origin or <select
  object>: pick
Specify second extension line origin: pick
Specify dimension line location or [Mtext/
Text/Angle/Horizontal/Vertical/Rotated]: pick
Dimension text = 180
Command:
```

Fig. 6.4 shows the 180 dimension. Follow exactly the same procedure for the 110 dimension.

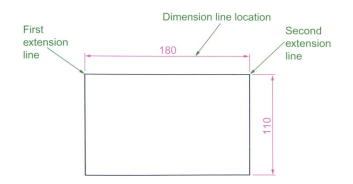

Fig. 6.4 First example – **Linear** dimension

Notes

1. If necessary use **Osnaps** to locate the extension line locations.

2. At the prompt:

```
Specify first extension line origin or [select
  object]:
```

    Also allows the line being dimensioned to be *picked*.

3. The drop-down menu from the **Line** tool icon contains the following tool icons – **Angular**, **Linear**, **Aligned**, **Arc Length**, **Radius**, **Diameter**, **Jog Line** and **Ordinate**. Refer to Fig. 6.1 when working through the examples below. **Note** – when a tool is chosen from this menu, the icon in the panel changes to the selected tool icon.

Second example – Aligned Dimension (Fig. 6.5)

1. Construct the outline Fig. 6.5 using the **Line** tool.

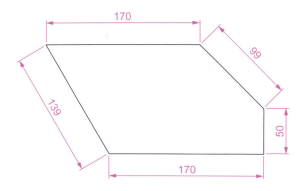

Fig. 6.5 Second example – **Aligned** dimension

2. Make the **Dimensions** layer current (**Home/Layers** panel).
3. *Left-click* the **Aligned** tool icon (see Fig. 6.1) and dimension the outline. The prompts and replies are similar to the first example.

Third example – Radius Dimension (Fig. 6.6)

1. Construct the outline Fig. 6.5 using the **Line** and **Fillet** tools.
2. Make the **Dimensions** layer current (**Home/Layers** panel).
3. *Left-click* the **Radius** tool icon (see Fig. 6.1). The command line shows:

```
Command:_dimradius
Select arc or circle: pick one of the arcs
Dimension text = 30
Specify dimension line location or [Mtext/Text/
  Angle]: pick
Command:
```

4. Continue dimensioning the outline as shown in Fig. 6.6.

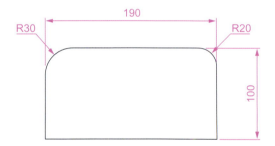

Fig. 6.6 Third example – **Radius** dimension

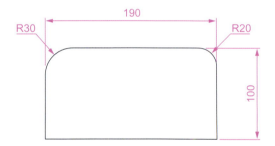

Notes

1. At the prompt:
   ```
   [Mtext/Text/Angle]:
   ```
   If a **t** (Text) is *entered*, another number can be *entered*, but remember if the dimension is a radius the letter **R** must be *entered* as a prefix to the new number.

2. If the response is **a** (Angle), and an angle number is *entered* the text for the dimension will appear at an angle.

3. If the response is **m** (Mtext) the **Text Formatting** dialog appears together with a box in which new text can be *entered*. See page 147.

4. Dimensions added to a drawing using other tools from the **Annotate/Dimensions** panel should be practised.

CHAPTER 6

Adding dimensions from the command line

From Figs 6.1 and 6.2 it will be seen that there are some dimension tools which have not been described in examples. Some operators may prefer *entering* dimensions from the command line. This involves abbreviations for the required dimension such as:

For **Linear Dimension** – **hor** (horizontal) or **ve** (vertical);
For **Aligned Dimension** – **al**;
For **Radius Dimension** – **ra**;
For **Diameter Dimension** – **d**;
For **Angular Dimension** – **an**;
For **Dimension Text Edit** – **te**;
For **Quick Leader** – **l**.

And to exit from the dimension commands – **e** (Exit).

First example – hor and ve (horizontal and vertical) – Fig. 6.8

1. Construct the outline Fig. 6.7 using the **Line** tool. Its dimensions are shown in Fig. 6.8.

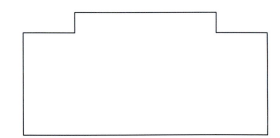

Fig. 6.7 First example – outline to dimension

2. Make the **Dimensions** layer current (**Home/Layers** panel).
3. At the command line *enter* **dim**. The command line will show:

```
Command: enter dim right-click
Dim: enter hor (horizontal) right-click
Specify first extension line origin or <select
  object>: pick
Specify second extension line origin: pick
Non-associative dimension created.
Specify dimension line location or [Mtext/Text/
  Angle]: pick
Enter dimension text <50>: right-click
```

```
Dim: right-click
HORIZONTAL
Specify first extension line origin or <select
  object>: pick
Specify second extension line origin: pick
Non-associative dimension created.
Specify dimension line location or [Mtext/Text/
  Angle/Horizontal/Vertical/Rotated]: pick
Enter dimension text <140>: right-click
Dim: right-click
```

And the 50 and 140 horizontal dimensions are added to the outline.

4. Continue to add the right-hand 50 dimension. Then when the command line shows:

```
Dim: enter ve (vertical) right-click
Specify first extension line origin or <select
  object>: pick
Specify second extension line origin: pick
Specify dimension line location or [Mtext/Text/
  Angle/Horizontal/Vertical/Rotated]: pick
Dimension text <20>: right-click
Dim: right-click
VERTICAL
Specify first extension line origin or <select
  object>: pick
Specify second extension line origin: pick
Specify dimension line location or [Mtext/Text/
  Angle/Horizontal/Vertical/Rotated]: pick
Dimension text <100>: right-click
Dim: enter e (Exit) right-click
Command:
```

The result is shown in Fig. 6.8.

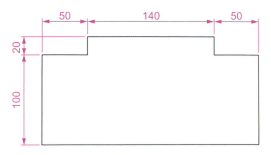

Fig. 6.8 First example – horizontal and vertical dimensions

Second example – an (Angular) – Fig. 6.10

1. Construct the outline Fig. 6.9 – a pline of width = 1.

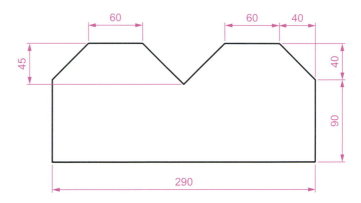

Fig. 6.9 Second example – outline for dimensions

2. Make the **Dimensions** layer current (**Home/Layers** panel).

3. At the command line:

```
Command: enter dim right-click
Dim: enter an right-click
Select arc, circle, line or <specify vertex>: pick
Select second line: pick
Specify dimension arc line location or [Mtext/
  Text/Angle/Quadrant]: pick
Enter dimension <90>: right-click
Enter text location (or press ENTER): pick
Dim:
```

And so on to add the other angular dimensions.

The result is given in Fig. 6.10.

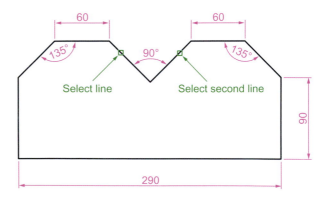

Fig. 6.10 Second example – **an** (Angular) dimension

Third example – I (Leader) – Fig. 6.12

1. Construct Fig. 6.11.

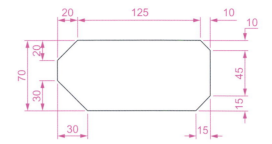

Fig. 6.11 Third example – outline for dimensioning

2. Make the **Dimensions** layer current (**Home/Layers** panel).
3. At the command line:

```
Command: enter dim right-click
Dim: enter l (Leader) right-click
Leader start: enter nea (osnap nearest)
  right-click to pick one of the chamfer
  lines
To point: pick
To point: pick
To point: right-click
Dimension text <0>: enter CHA 10 × 10 right-click
Dim: right-click
```

Continue to add the other leader dimensions – Fig. 6.12.

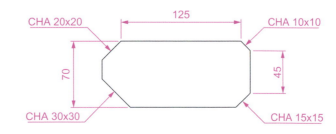

Fig. 6.12 Third example – I (Leader) dimensions

CHAPTER 6

Fourth example – te (dimension text edit) – Fig. 6.14

1. Construct Fig. 6.13.

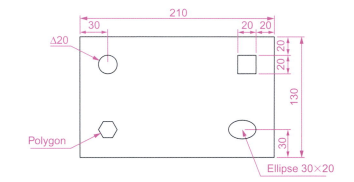

Fig. 6.13 Fourth example – dimensioned drawing

2. Make the **Dimensions** layer current (**Home/Layers** panel).
3. At the command line:

```
Command: enter dim right-click
Dim: enter te (tedit) right-click
Select dimension: pick the dimension to be changed
Specify new location for text or [Left/Right/
  Center/Home/Angle]: either pick or enter a prompt
  capital letter
Dim:
```

The results as given in Fig. 6.14 show dimensions which have been moved. The **210** dimension changed to the left-hand end of the dimension line, the **130** dimension changed to the left-hand end of the dimension line and the **30** dimension position changed.

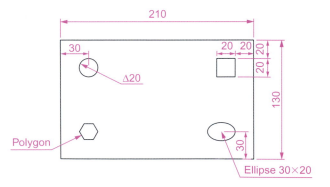

Fig. 6.14 Fourth example – dimensions amended with **tedit**

The Arc Length tool (**Fig. 6.15**)

1. Construct two arcs of different sizes as in Fig. 6.15.

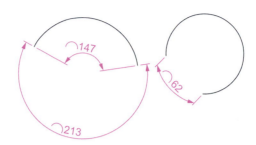

Fig. 6.15 Examples – **Arc Length** tool

2. Make the **Dimensions** layer current (**Home/Layers** panel).
3. Call the **Arc Length** tool from the **Annotate/Dimensions** panel (see Fig. 6.3) or *enter* **dimarc** at the command line. The command line shows:

```
Command: _dimarc
Select arc or polyline arc segment: pick an arc
Specify arc length dimension location, or [Mtext/
  Text/Angle/Partial/Leader]: pick a suitable
  position
Dimension text = 147
Command:
```

Examples on two arcs are shown in Fig. 6.15.

CHAPTER 6

The Jogged tool (**Fig. 6.16**)

1. Draw a circle and an arc as indicated in Fig. 6.16.

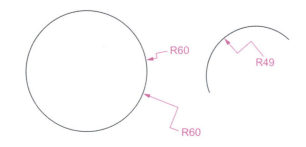

Fig. 6.16 Examples – the **Jogged** tool

2. Make the **Dimensions** layer current (**Home/Layers** panel).
3. Call the **Jogged** tool, either with a *left-click* on its tool icon in the
   Annotation/Dimension panel (see Fig. 6.1) or by *entering* **jog** at the
   command line. The command line shows:

```
Command: _dimjogged
Select arc or circle: pick the circle or the
 arc
Specify center location override:
 pick
Dimension text = 60
Specify dimension line location or [Mtext/Text/
 Angle]: pick
Specify jog location: pick
Command:
```

The results of placing as jogged dimension on a circle and an arc are
shown in Fig. 6.16.

Dimension tolerances

Before simple tolerances can be included with dimensions, new settings
will need to be made in the **Dimension Style Manager** dialog as
follows:

1. Open the dialog. The quickest way of doing this is to *enter* **d** at the
   command line followed by a *right-click*. This opens up the dialog.
2. *Click* the **Modify…** button of the dialog, followed by a *left-click* on
   the **Primary Units** tab and in the resulting sub-dialog make settings
   as shown in Fig. 6.17. Note the changes in the preview box of the
   dialog.

Example – tolerances (Fig. 6.19)

1. Construct the outline Fig. 6.18.
2. Make the **Dimensions** layer current (**Home/Layers** panel).
3. Dimension the drawing using either tools from the **Dimension** panel
   or by *entering* abbreviations at the command line. Because tolerances
   have been set in the **Dimension Style Manager** dialog (Fig. 6.17), the
   toleranced dimensions will automatically be added to the drawing
   (Fig. 6.19).

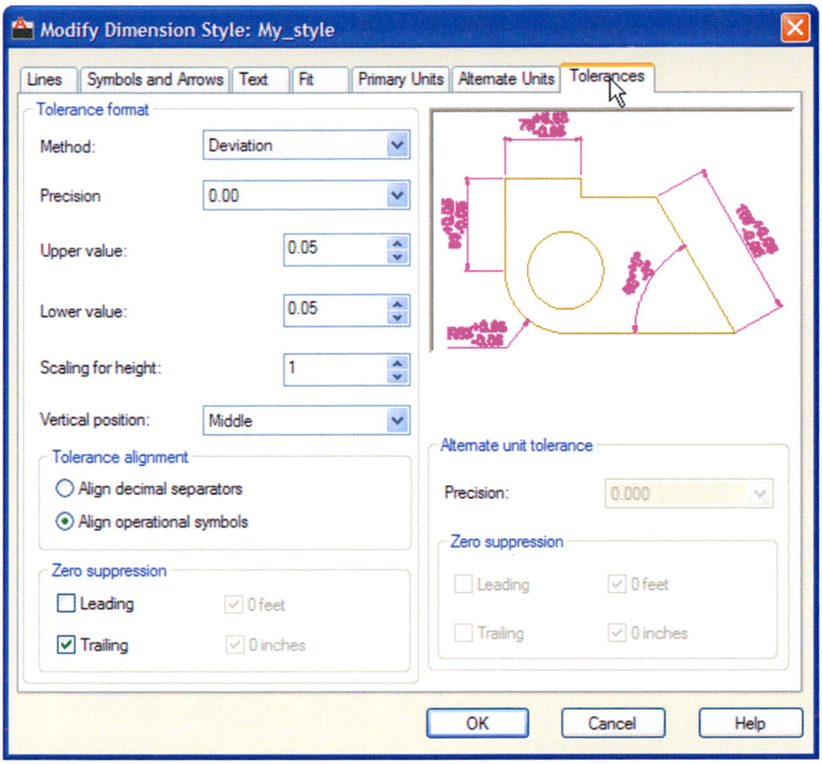

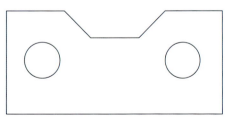

Fig. 6.17 The **Tolerances** sub-dialog of the **Modify Dimension Style** dialog

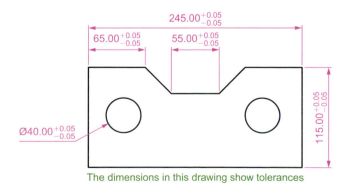

Fig. 6.18 First example – simple tolerances – outline

$$245.00^{+0.05}_{-0.05}$$
$$65.00^{+0.05}_{-0.05}$$ $$55.00^{+0.05}_{-0.05}$$
$$115.00^{+0.05}_{-0.05}$$
$$\varnothing 40.00^{+0.05}_{-0.05}$$

The dimensions in this drawing show tolerances

Fig. 6.19 Example – tolerances

Text

There are two main methods of adding text to drawings – **Multiline Text** and **Single Line Text**.

Example – Single Line Text (Fig. 6.19)

1. Open the drawing from the example on tolerances – Fig. 6.19.
2. Make the **Text** layer current (**Home/Layers** panel).
3. At the command line *enter* **dt** (for Single Line Text) followed by a *right-click*:

```
Command: enter dt right-click
TEXT
Current text style "ARIAL" Text height: 8
  Annotative No:
Specify start point of text or [Justify/Style]:
  pick
Specify rotation angle of text <0>: right-click
Enter text: enter The dimensions in this drawing
  show tolerances press the Return key twice
Command:
```

The result is given in Fig. 6.19.

Notes

1. When using **Dynamic Text** the **Return** key of the keyboard is pressed when the text has been *entered*. A *right-click* does not work.

2. At the prompt:

```
Specify start point of text or [Justify/Style]:
  enter s (Style) right-click
Enter style name or [?] <ARIAL>: enter ?
  right-click
Enter text style(s) to list <*>: right-click
```

And an **AutoCAD Text Window** (Fig. 6.20) appears listing all the styles which have been selected in the **Text Style** dialog (see page 145).

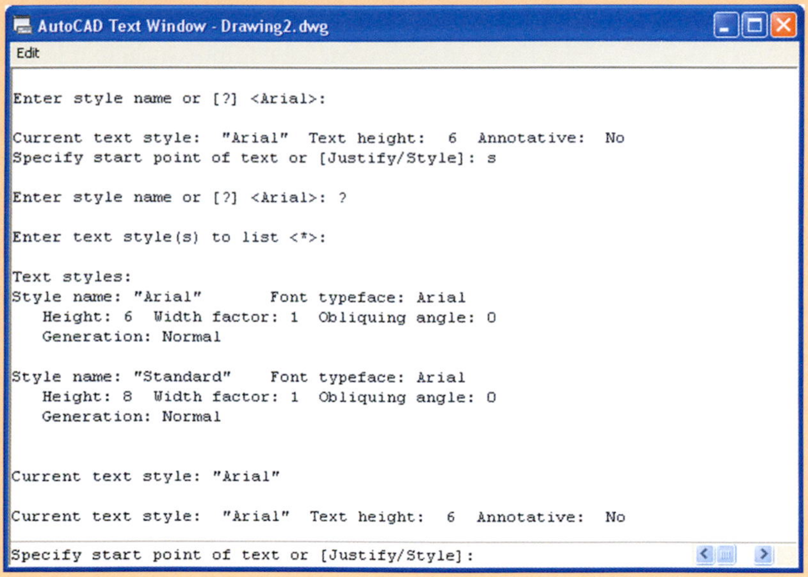

Fig. 6.20 The **AutoCAD Text Window**

3. In order to select the required text style its name must be *entered* at the prompt:

```
Enter style name or [?] <ARIAL>: enter Romand
    right-click
```

And the text *entered* will be in the **Romand** style of height **9**. But only if that style was previously been selected in the **Text Style** dialog.

4. Fig. 6.21 shows some text styles from the **AutoCAD Text Window**.

This is the TIMES text

This is ROMANC text

This is ROMAND text

This is STANDARD text

This is ITALIC text

This is ARIAL text

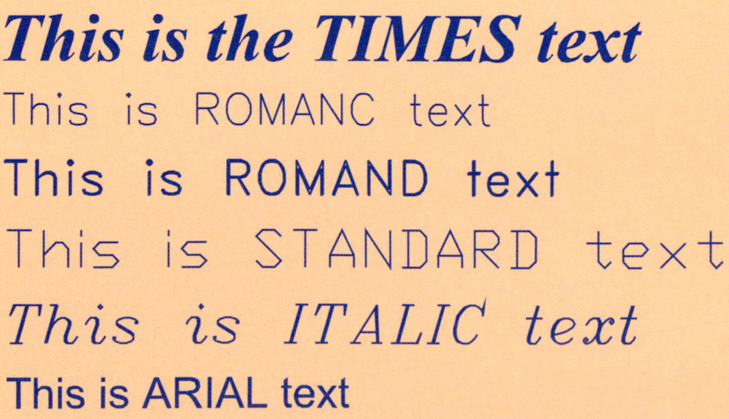

Fig. 6.21 Some text fonts

CHAPTER 6

5. There are two types of text fonts available in AutoCAD 2011 – the **5**. There are two types of text fonts available in AutoCAD 2011 – the **AutoCAD SHX** fonts and the **Windows True Type** fonts. The styles shown in Fig. 6.21 are the **ITALIC**, **ROMAND**, **ROMANS** and **STANDARD** styles are AutoCAD text fonts. The **TIMES** and **ARIAL** styles are **Windows True Type** styles. Most of the **True Type** fonts can be *entered* in **Bold**, **Bold Italic**, **Italic** or **Regular** styles, but these variations are not possible with the AutoCAD fonts.

6. The **Font name** popup list of the **Text Style** dialog shows that a large number of text styles are available to the AutoCAD 2011 operator. It is advisable to practise using a variety of these fonts to familiarise oneself with the text opportunities available with AutoCAD 2011.

Example – Multiline Text (Fig. 6.23)

1. Make the **Text** layer current (**Home/Layers** panel).
2. Either *left-click* on the **Multiline Text** tool icon in the **Annotate/Text** panel (Fig. 6.22) or *enter* **t** at the command line:

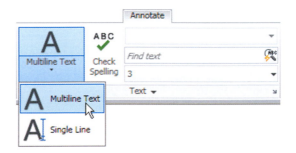

Fig. 6.22 Selecting **Multiline Text…** from the **Annotate/Text** panel

```
Command:_mtext
Current text style: "Arial" Text height: 6
  Annotative No
Specify first corner: pick
Specify opposite corner or [Height/Justify/Line
  spacing/Rotation/Style/Width/Columns]: pick
```

As soon as the **opposite corner** is *picked*, the **Text Formatting** box appears (Fig. 6.23). Text can now be *entered* as required within the box as indicated in this illustration.

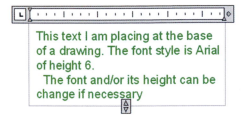

Fig. 6.23 Example – **Multiline Text** *entered* in the text box

When all the required text has been entered *left-click* and the text box disappears leaving the text on screen.

Symbols used in text

When text has to be added by *entering* letters and figures as part of a dimension, the following symbols must be used:

To obtain **Ø75** *enter* **%%c75**;
To obtain **55%** *enter* **55%%%**;
To obtain **±0.05** *enter* **%%p0.05**;
To obtain **90°** *enter* **90%%d**.

Checking spelling

> **Note**
>
> When a misspelt word or a word not in the AutoCAD spelling dictionary is *entered* in the **Multiline Text** box, red dots appear under the word, allowing immediate correction.
>
> There are two methods for the checking of spelling in AutoCAD 2011.

First example – spell checking – ddedit (Fig. 6.24)

1. *Enter* some badly spelt text as indicated in Fig. 6.24.
2. *Enter* **ddedit** at the command line.
3. *Left-click* on the text. The text is highlighted. Edit the text as if working in a word processing application and when satisfied *left-click* followed by a *right-click*.

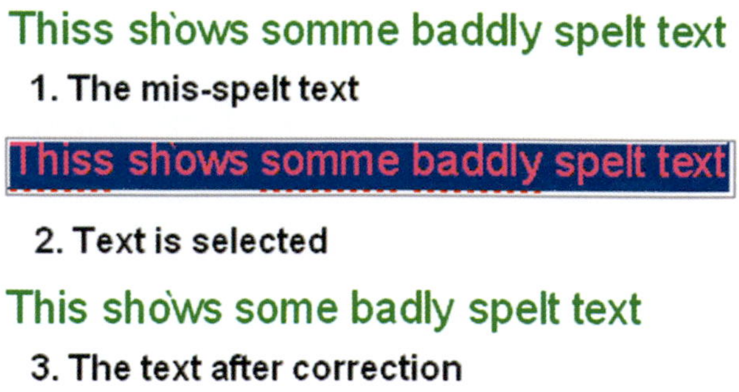

Thiss shows somme baddly spelt text

1. The mis-spelt text

Thiss shows somme baddly spelt text

2. Text is selected

This shows some badly spelt text

3. The text after correction

Fig. 6.24 First example – **spell checking – ddedit**

Second example – the Spelling tool (Fig. 6.25)

1. *Enter* some badly spelt text as indicated in Fig. 6.25.
2. Either *click* the **Spell Check…** icon in the **Annotate/Text** panel (Fig. 6.26) or *enter* **spell** or **sp** at the command line.
3. The **Check Spelling** dialog appears (Fig. 6.25). In the **Where to look** field select **Entire drawing** from the field's popup list. The first badly spelt word is highlighted with words to replace them listed in the **Suggestions** field. Select the appropriate correct spelling as shown.

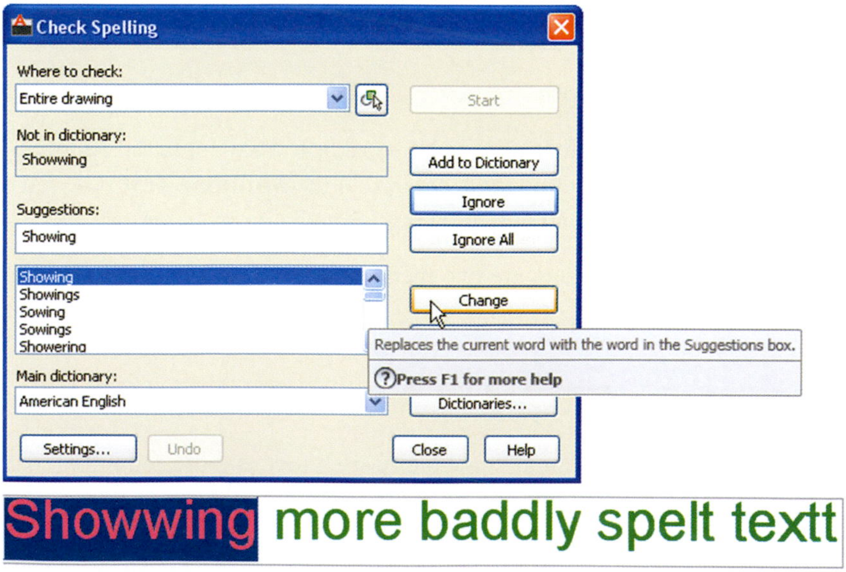

Showwing more baddly spelt textt

Fig. 6.25 Second example – the **Check Spelling** dialog

Fig. 6.27 The
AutoCAD Message
window showing
that spelling check is
complete

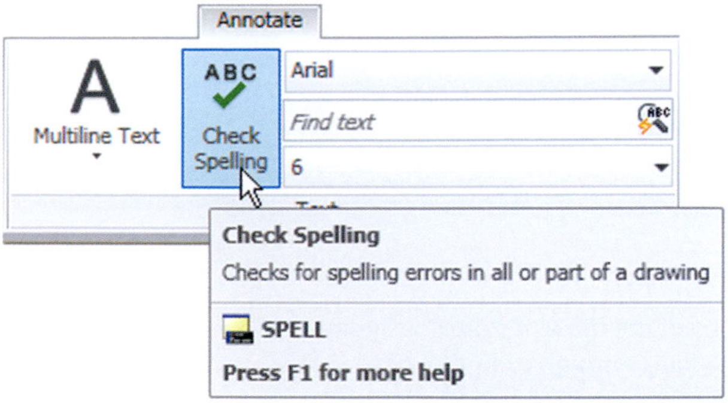

Fig. 6.26 The **Spell Check...** icon in the **Annotate/Text** panel

Continue until all text is checked. When completely checked an
AutoCAD Message appears (Fig. 6.27). If satisfied *click* its
OK button.

REVISION NOTES

1. In the Line and Arrows sub-dialog of the Dimension Style Manager dialog Lineweights were set to 0.3. If these lineweights are to show in the drawing area of AutoCAD 2011, the Show/Hide Lineweight button in the status bar must be set ON.
2. Dimensions can be added to drawings using the tools from the Annotate/Dimensions panel or by *entering* dim, followed by abbreviations for the tools at the command line.
3. It is usually advisable to use osnaps when locating points on a drawing for dimensioning.
4. The Style and Angle of the text associated with dimensions can be changed during the dimensioning process.
5. When wishing to add tolerances to dimensions it will probably be necessary to make new settings in the Dimension Style Manager dialog.
6. There are two methods for adding text to a drawing – Single Line Text and Multiline Text.
7. When adding Single Line Text to a drawing, the Return key must be used and not the right-hand mouse button.
8. Text styles can be changed during the process of adding text to drawings.
9. AutoCAD 2011 uses two types of text style – AutoCAD SHX fonts and Windows True Type fonts.
10. Most True Type fonts can be in bold, bold italic, italic or regular format. AutoCAD fonts can only be added in the single format.
11. To obtain the symbols Ø; $\pm$; °; % use %%c; %%p; %%d; %%% before the figures of the dimension.
12. Text spelling can be checked with by selecting Object/Text/Edit... from the Modify drop-down menu, by selecting Spell Check... from the Annotate/Text panel, or by entering spell or sp at the command line.

CHAPTER 6

Exercises

Methods of constructing answers to the following exercises can be found in the free website:

http://books.elsevier.com/companions/978-0-08-096575-8

1. Open any of the drawings previously saved from working through examples or as answers to exercises and add appropriate dimensions.

2. Construct the drawing Fig. 6.28 but in place of the given dimensions add dimensions showing tolerances of 0.25 above and below.

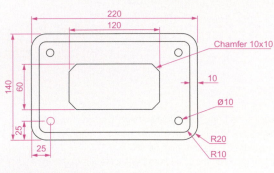

Fig. 6.28 Exercise 2

3. Construct and dimension the drawing Fig. 6.29.

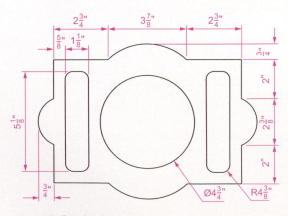

Fig. 6.29 Exercise 3

4. Construct two polygons as in Fig. 6.30 and add all diagonals. Set osnaps **endpoint** and **intersection** and using the lines as in Fig. 6.30 construct the stars as shown using a polyline of Width = 3. Next erase all unwanted lines. Dimension the angles labelled **A**, **B**, **C** and **D**.

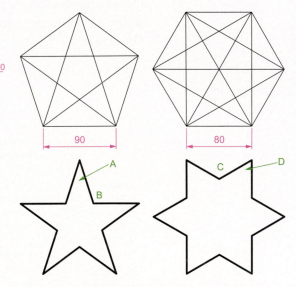

Fig. 6.30 Exercise 4

5. Using the text style **Arial** of height **20** and enclosing the wording within a pline rectangle of Width = 5 and Fillet = 10, construct Fig. 6.31.

Fig. 6.31 Exercise 5

Chapter 7

Orthographic and isometric

AIM OF THIS CHAPTER

The aim of this chapter is to introduce methods of constructing views in orthographic projection and the construction of isometric drawings.

151

Orthographic projection

Orthographic projection involves viewing an article being described in a technical drawing from different directions – from the front, from a side, from above, from below or from any other viewing position. Orthographic projection often involves:

The drawing of details which are hidden, using hidden detail lines.
Sectional views in which the article being drawn is imagined as being cut through and the cut surface drawn.
Centre lines through arcs, circles spheres and cylindrical shapes.

An example of an orthographic projection

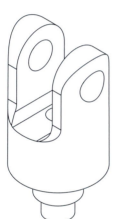

Fig. 7.1 Example – orthographic projection – the solid being drawn

Taking the solid shown in Fig. 7.1 – to construct a three-view orthographic projection of the solid:

1. Draw what is seen when the solid is viewed from its left-hand side and regard this as the **front** of the solid. What is drawn will be a **front view** (Fig. 7.2).

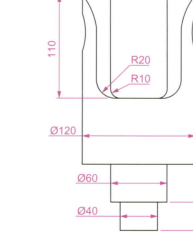

Fig. 7.2 The **front view** of the solid

2. Draw what is seen when the solid is viewed from the left-hand end of the front view. This produces an **end view**. Fig. 7.3 shows the end view alongside the front view.
3. Draw what is seen when the solid is viewed from above the front view. This produces a **plan**. Fig. 7.4 shows the plan below the front view.
4. In the **Home/Layers** panel in the **Layer** list *click* on **Centre** to make it the current layer (Fig. 7.5). All lines will now be drawn as centre lines.

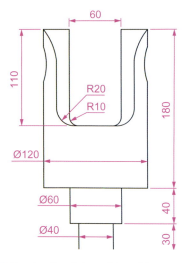
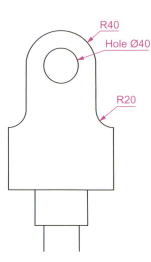

Fig. 7.3 **Front** and **end** views of the solid

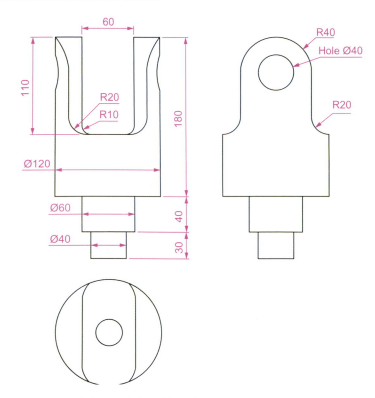

Fig. 7.4 **Front** and **end** views and **plan** of the solid

5. In the three-view drawing add centre lines.
6. Make the **Hidden** layer the current layer and add hidden detail lines.
7. Make the **Text** layer current and add border lines and a title block.
8. Make the **Dimensions** layer current and add all dimensions.

The completed drawing is shown in Fig. 7.6.

CHAPTER 7

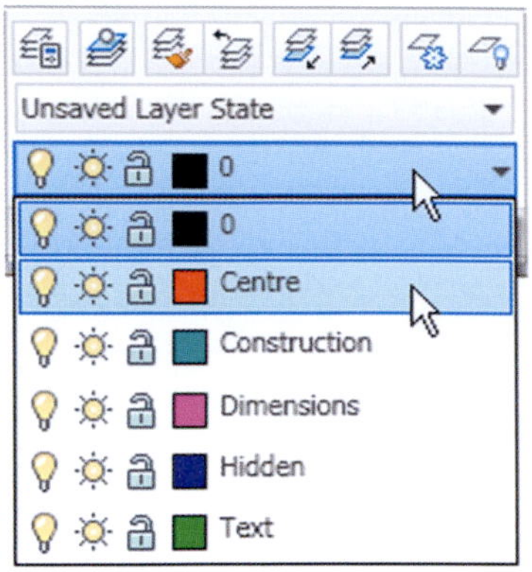

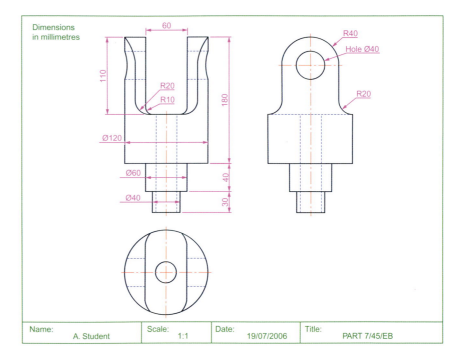

Fig. 7.5 Making the layer **Centre** current from the **Home/Layers** panel

Fig. 7.6 The completed working drawing of the solid

First angle and third angle

There are two types of orthographic projection – **first angle** and **third angle**. Fig. 7.7 is a pictorial drawing of the solid used to demonstrate the two angles. Fig. 7.8 shows a three-view **first angle projection** and Fig. 7.9 the same views in **third angle**.

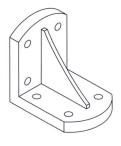

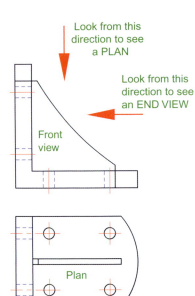

Look from this
direction to see
a PLAN

Look from this
direction to see
an END VIEW

Front
view

End view

Plan

Fig. 7.7 The solid
used to demonstrate
first and third angles of
projection

Fig. 7.8 A **first angle** projection

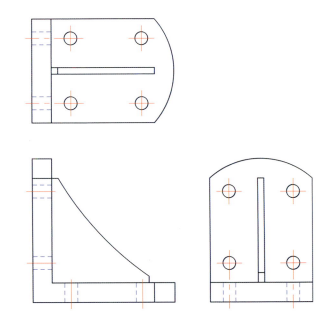

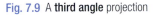

Fig. 7.9 A **third angle** projection

In both angles the viewing is from the same directions. The
difference is that the view as seen is placed on the viewing side of
the front view in **third angle** and on the opposite side to the viewing in
first angle.

CHAPTER 7

Adding hatching

In order to show internal shapes of a solid being drawn in orthographic projection, the solid is imagined as being cut along a plane and the cut surface then drawn as seen. Common practice is to **hatch** the areas which then show in the cut surface. Note the section plane line, the section label and the hatching in the sectional view (Fig. 7.10).

To add the hatching as shown in Fig. 7.10:

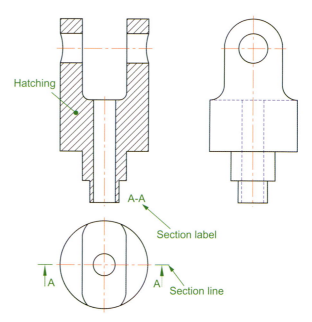

Fig. 7.10 A sectional view

1. Call the **Hatch** tool with a *left-click* on its tool icon in the **Home/Draw** panel (Fig. 7.11). A new tab **Hatch Creation** is created and opens the **Hatch Creation** ribbon (Fig. 7.12), but only if the ribbon is active.

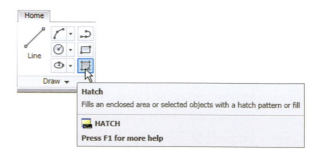

Fig. 7.11 The **Hatch** tool icon and tooltip from the **Home/Draw** panel

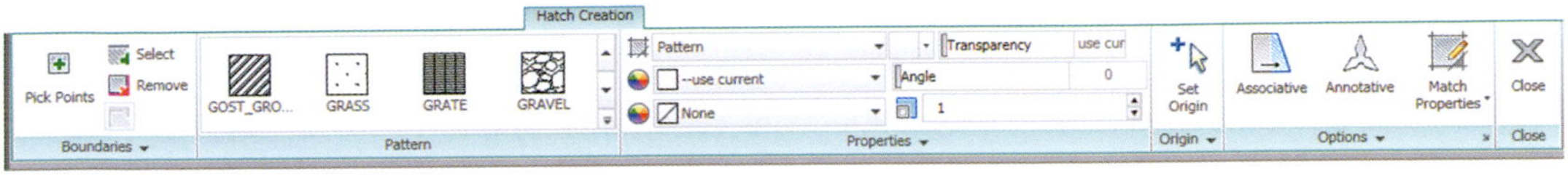

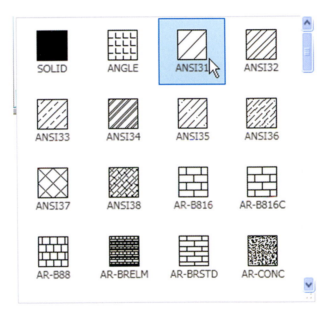

Fig. 7.12 The **Hatch Creation** tab and ribbon

2. In the **Hatch Creation/Pattern** panel *click* the bottom arrow on the right of the panel and from the palette which appears *pick* the **ANI31** pattern (Fig. 7.13).
3. In the **Hatch Creation/Properties** panel adjust the **Hatch Scale** to **2** (Fig. 7.14).

Fig. 7.13 Selecting **ANSI31** pattern from the **Hatch Creation/Pattern** panel

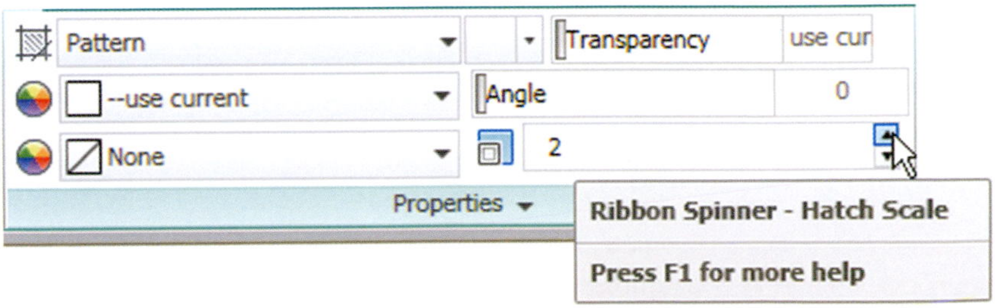

Fig. 7.14 Setting the **Hatch Scale to** 2 in the **Hatch Creation/Properties** panel

CHAPTER 7

4. In the **Hatch Creation/Boundaries** panel *left-click* the **Pick Points** icon (Fig. 7.15).

5. *Pick* the points in the front view (left-hand drawing of Fig. 7.16) and the *picked* points hatch. If satisfied the hatching is correct *right-click* (right-hand drawing of Fig. 7.16).

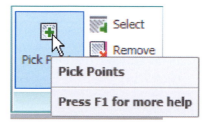

Fig. 7.15 Select **Pick Points** from the **Hatch Creation/Boundaries** panel

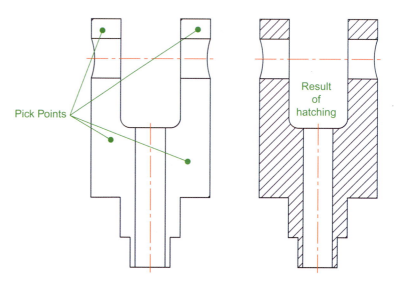

Fig. 7.16 The result of hatching

Isometric drawing

> **Note**
>
> Isometric drawing must not be confused with solid model drawing, examples of which are given in Chapters 12–19. Isometric drawing is a 2D method of describing objects in a pictorial form.

Setting the AutoCAD window for isometric drawing

To set the AutoCAD 2011 window for the construction of isometric drawings:

1. At the command line:

```
Command: enter snap
Specify snap spacing or [On/Off/Aspect/Rotate/
  Style/Type] <5>: s (Style)
Enter snap grid style [Standard/Isometric] <S>:
  i (Isometric)
Specify vertical spacing <5>: right-click
Command:
```

And the grid dots in the window assume an isometric pattern as shown in Fig. 7.17. Note also the cursor hair lines which are at set in an **Isometric Left** angle.

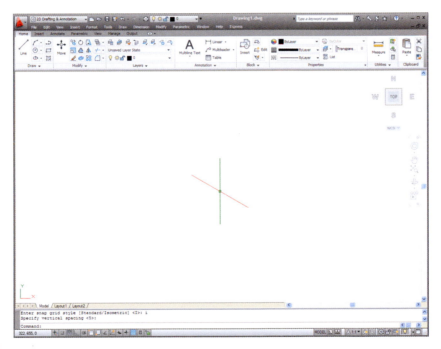

Fig. 7.17 The AutoCAD grid points set for isometric drawing

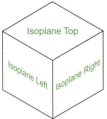

Fig. 7.18 The three isoplanes

2. There are three isometric angles – **Isoplane Top**, **Isoplane Left** and **Isoplane Right**. These can be set by pressing either the **F5** function key or the **Ctrl** and **E** keys. Repeated pressing of either of these 'toggles' between the three settings. Fig. 7.18 is an isometric view showing the three isometric planes.

CHAPTER 7

The isometric circle

Circles in an isometric drawing show as ellipses. To add an isometric circle to an isometric drawing, call the **Ellipse** tool. The command line shows:

```
Command: _ellipse
Specify axis endpoint of ellipse or [Arc/Center/
    Isocircle]: enter i (Isocircle) right-click
Specify center of isocircle: pick or enter
    coordinates
Specify radius of isocircle or [Diameter]: enter a
    number
Command:
```

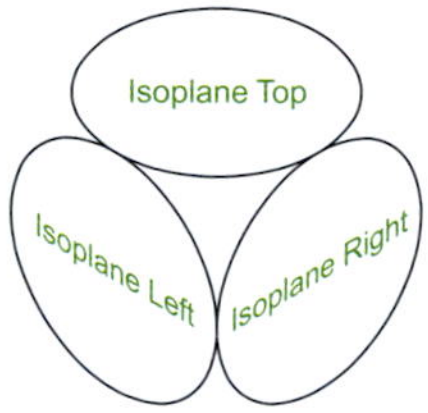

And the isocircle appears. Its isoplane position is determined by which of the isoplanes is in operation at the time the isocircle was formed. Fig. 7.19 shows these three isoplanes containing isocircles.

Fig. 7.19 The three isocircles

Examples of isometric drawings

First example – isometric drawing (Fig. 7.22)

1. This example is to construct an isometric drawing to the details given in the orthographic projection (Fig. 7.20). Set Snap on (press the **F9** function key) and Grid on (**F7**).
2. Set Snap to Isometric and set the isoplane to **Isoplane Top** using **F5**.
3. With **Line**, construct the outline of the top of the model (Fig. 7.19) working to the dimensions given in Fig. 7.18.
4. Call **Ellipse** tool and set to isocircle and add the isocircle of radius 20 centred in its correct position in the outline of the top (Fig. 7.21).
5. Set the isoplane to Isoplane Right and with the **Copy** tool, copy the top with its ellipse vertically downwards 3 times as shown in Fig. 7.22.
6. Add lines as shown in Fig. 7.21.
7. Finally using **Trim** remove unwanted parts of lines and ellipses to produce Fig. 7.22.

Second example – isometric drawing (Fig. 7.24)

Fig. 7.23 is an orthographic projection of the model of which the isometric drawing is to be constructed. Fig. 7.24 shows the stages in its construction. The numbers refer to the items in the list below:

1. In **Isoplane Right** construct two isocircles of radii 10 and 20.
2. Add lines as in drawing **2** and trim unwanted parts of isocircle.

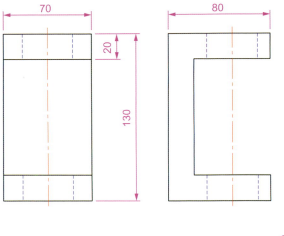

Fig. 7.20 First example – isometric drawing – the model

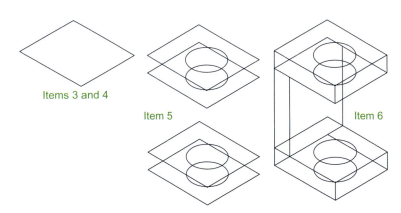

Items 3 and 4

Item 5

Item 6

Fig. 7.21 First example – isometric drawing – items **3**, **4**, **5** and **6**

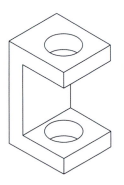

Fig. 7.22 First example – isometric drawing

3. With **Copy** copy 3 times as in drawing **3**.
4. With **Trim** trim unwanted lines and parts of isocircle (drawing **4**).
5. In **Isoplane Left** add lines as in drawing **5**.
6. In **Isoplane Right** add lines and isocircles as in drawing **6**.
7. With **Trim** trim unwanted lines and parts of isocircles to complete the isometric drawing – drawing **7**.

CHAPTER 7

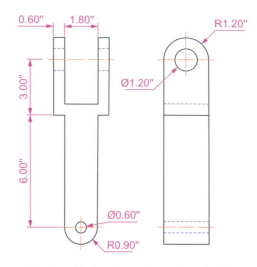

Fig. 7.23 Second example – isometric drawing – orthographic projection

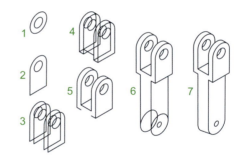

Fig. 7.24 Second example – isometric drawing – stages in the construction

REVISION NOTES

1. There are, in the main, two types of orthographic projection – first angle and third angle.
2. The number of views included in an orthographic projection depends upon the complexity of the component being drawn – a good rule to follow is to attempt fully describing the object in as few views as possible.
3. Sectional views allow parts of an object which are normally hidden from view to be more fully described in a projection.
4. When a layer is turned OFF, all constructions on that layer disappear from the screen.
5. Frozen layers cannot be selected, but note that layer 0 cannot be frozen.
6. Isometric drawing is a 2D pictorial method of producing illustrations showing objects. It is not a 3D method of showing a pictorial view.
7. When drawing ellipses in an isometric drawing the Isocircle prompt of the Ellipse tool command line sequence must be used.
8. When constructing an isometric drawing Snap must be set to Isometric mode before construction can commence.

Exercises

Methods of constructing answers to the following exercises can be found in the free website:

http://books.elsevier.com/companions/978-0-08-096575-8

Fig. 7.25 is an isometric drawing of a slider fitment on which the three exercises **1**, **2** and **3** are based.

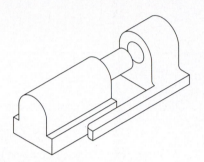

Fig. 7.25 Exercises 1, 2 and 3 – an isometric drawing of the three parts of the slider on which these exercises are based

1. Fig. 7.26 is a first angle orthographic projection of part of the fitment shown in the isometric drawing Fig. 7.23. Construct a three-view third angle orthographic projection of the part.

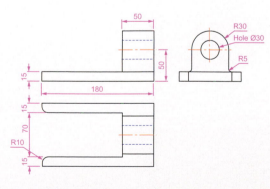

Fig. 7.26 Exercise 1

2. Fig. 7.27 is a first angle orthographic projection of the other part of the fitment. Construct a three-view third angle orthographic projection of the part.

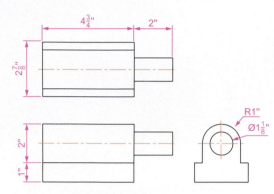

Fig. 7.27 Exercises 2 and 3

3. Construct an isometric drawing of the part shown in Fig. 7.27.

4. Construct a three-view orthographic projection in an angle of your own choice of the tool holder assembled as shown in the isometric drawing Fig. 7.28. Details are given in Fig. 7.29.

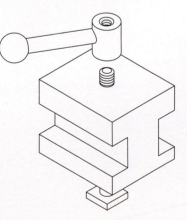

Fig. 7.28 Exercises 4 and 5 – orthographic projections of the three parts of the tool holder

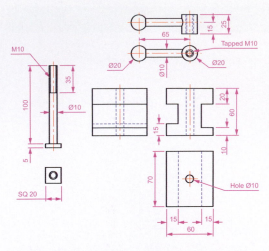

Fig. 7.29 Exercises 4 and 5 – orthographic drawing of the tool holder on which the two exercises are based

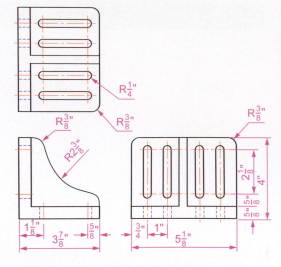

Fig. 7.31 Exercises 6 and 7 – an orthographic projection of the angle plate

5. Construct an isometric drawing of the body of the tool holder shown in Figs 7.28 and 7.29.

6. Construct the orthographic projection given in Fig. 7.29.

7. Construct an isometric drawing of the angle plate shown in Figs 7.30 and 7.31.

9. Construct the isometric drawing shown in Fig. 7.32 working to the dimensions given in Fig. 7.33.

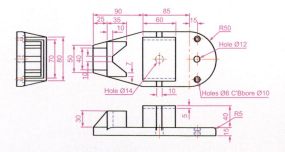

Fig. 7.32 Exercises 8 and 9

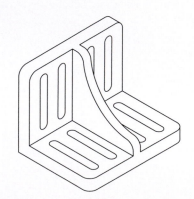

Fig. 7.30 An isometric drawing of the angle plate on which exercises 6 and 7 are based

8. Construct a third angle projection of the component shown in the isometric drawing Fig. 7.32 and the three-view first angle projection Fig. 7.33.

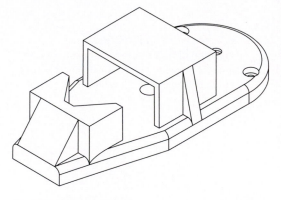

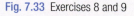

Fig. 7.33 Exercises 8 and 9

Hatching

AIM OF THIS CHAPTER

The aim of this chapter is to give further examples of the use of hatching in its various forms.

Introduction

In Chapter 7 an example of hatching of a sectional view in an orthographic projection was given. Further examples of hatching will be described in this chapter.

There are a large number of hatch patterns available when hatching drawings in AutoCAD 2011. Some examples from hatch patterns are shown in Fig. 8.1.

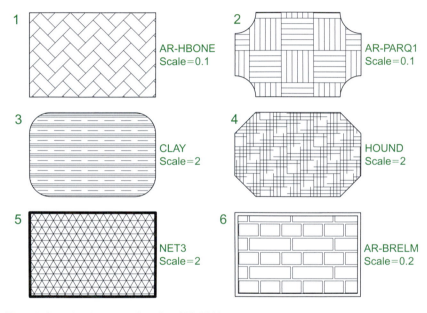

1 AR-HBONE Scale=0.1
2 AR-PARQ1 Scale=0.1
3 CLAY Scale=2
4 HOUND Scale=2
5 NET3 Scale=2
6 AR-BRELM Scale=0.2

Fig. 8.1 Some hatch patterns from AutoCAD 2011

Other hatch patterns can be selected from **Hatch Creation/Properties** panel, or the operator can design his/her own hatch patterns as **User Defined** patterns (Fig. 8.2).

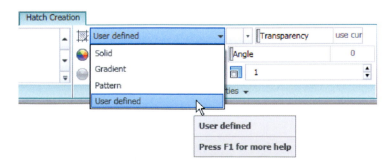

Fig. 8.2 The **User Defined** patterns in the **Hatch Creation/Properties** panel

First example – hatching a sectional view (Fig. 8.3)

Fig. 8.3 shows a two-view orthographic projection which includes a sectional end view. Note the following in the drawing:

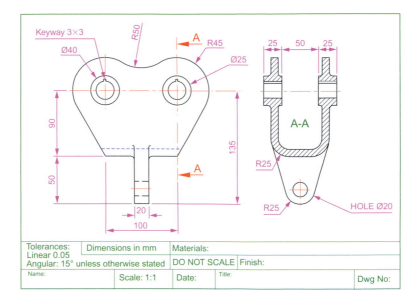

Fig. 8.3 First example – **Hatching**

1. The section plane line, consisting of a centre line with its ends marked **A** and arrows showing the direction of viewing to obtain the sectional view.
2. The sectional view labelled with the letters of the section plane line.
3. The cut surfaces of the sectional view hatched with the **ANSI31** hatch pattern, which is in general use for the hatching of engineering drawing sections.

Second example – hatching rules (Fig. 8.4)

Fig. 8.4 describes the stages in hatching a sectional end view of a lathe tool holder. Note the following in the section:

1. There are two angles of hatching to differentiate the separate parts of the section.
2. The section follows the general rule that parts such as screws, bolts, nuts, rivets, other cylindrical objects, webs and ribs, and other such features are shown as outside views within sections.

CHAPTER 8

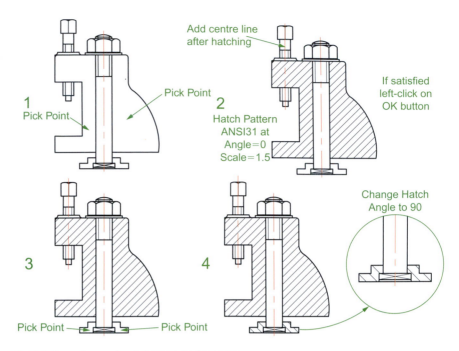

Fig. 8.4 Second example – hatching rules for sections

In order to hatch this example:

1. *Left-click* on the **Hatch** tool icon in the **Home/Draw** panel (Fig. 8.5). The ribbon changes to the **Hatch Creation** ribbon. *Entering* **hatch** or **h** at the command line has the same result.
2. *Left-click* **ANSI31** in the **Hatch Creation/Pattern** panel (Fig. 8.6).
3. Set the **Hatch Scale** to **1.5** in the **Hatch Creation/Properties** panel (Fig. 8.7).
4. *Left-click* **Pick Points** in the **Hatch Creation/Boundaries** panel and *pick* inside the areas to be hatched (Fig. 8.8).
5. The *picked* areas hatch. If satisfied with the hatching *right-click*. If not satisfied amend the settings and when satisfied *right-click*.

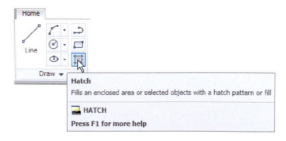

Fig. 8.5 *Left-click* on the **Hatch** tool icon in the **Home/Draw** panel

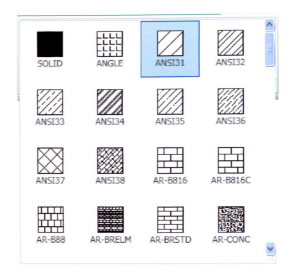

Fig. 8.6 Select **ANSI31** in the **Hatch Creation/Pattern** panel

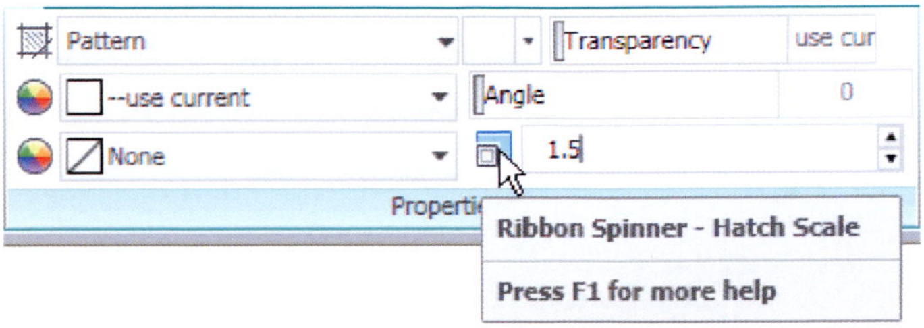

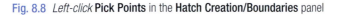

Fig. 8.7 Set the **Hatch Scale** in the **Hatch Creation/Properties** panel

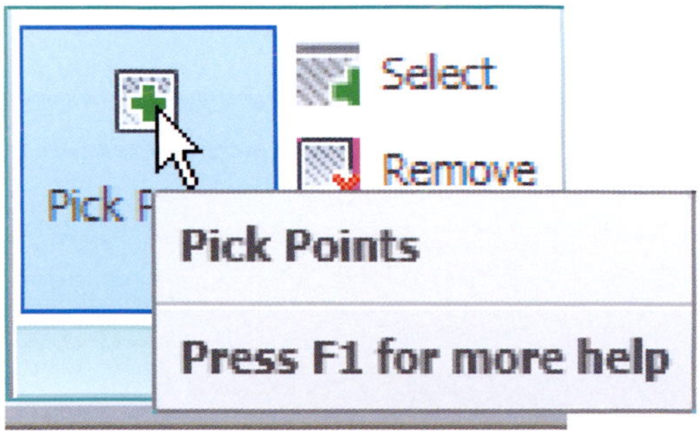

Fig. 8.8 *Left-click* **Pick Points** in the **Hatch Creation/Boundaries** panel

The Hatch and Gradient dialog

If the ribbon is not on screen, *entering* **hatch** or **h** at the command line brings the **Hatch and Gradient** dialog to screen (Fig. 8.9). The method of hatching given in the previous two examples is much the same whether

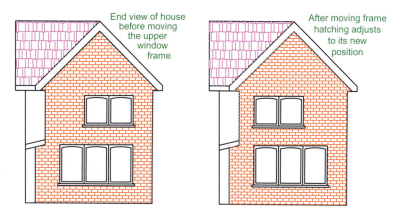

Fig. 8.9 The **Hatch and Gradient** dialog

using the tools in the **Hatch Creation** ribbon or using the **Hatch and Gradient** dialog. Fig. 8.9 shows the **ANSI Hatch Pattern** dialog and the **Pick Point**s button in the **Hatch and Gradient** dialog, which are *picked* for the same methods as described in the given examples.

Third example – Associative hatching (Fig. 8.10)

Fig. 8.10 shows two end view of a house. After constructing the left-hand view, it was found that the upper window had been placed in the wrong

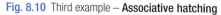

Fig. 8.10 Third example – **Associative hatching**

position. Using the **Move** tool, the window was moved to a new position. The brick hatching automatically adjusted to the new position. Such **associative hatching** is only possible if check box is **ON** – a tick in the check box in the **Options** area of the **Hatch and Gradient** dialog (Fig. 8.11).

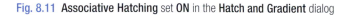

Fig. 8.11 Associative Hatching set **ON** in the **Hatch and Gradient** dialog

Fourth example – Colour gradient hatching (Fig. 8.12)

Fig. 8.12 shows two examples of hatching from the **Gradient** sub-dialog of the **Hatch and Gradient** dialog.

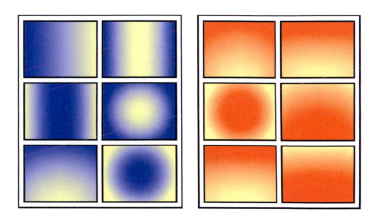

Fig. 8.12 Fourth example – Colour gradient hatching

1. Construct two outlines each consisting of six rectangles (Fig. 8.12).
2. *Click* **Gradient** in the drop-down menu in the **Hatch Creation/ Properties** panel (Fig. 8.13). In the **Hatch Creation/Pattern** panel which then appears, *pick* one of the gradient choices (Fig. 8.14), followed by a *click* in a single area of one of the rectangles in the left-hand drawing, followed by a *right-click*.
3. Repeat in each of the other rectangles of the left-hand drawing changing the pattern in each of the rectangles.

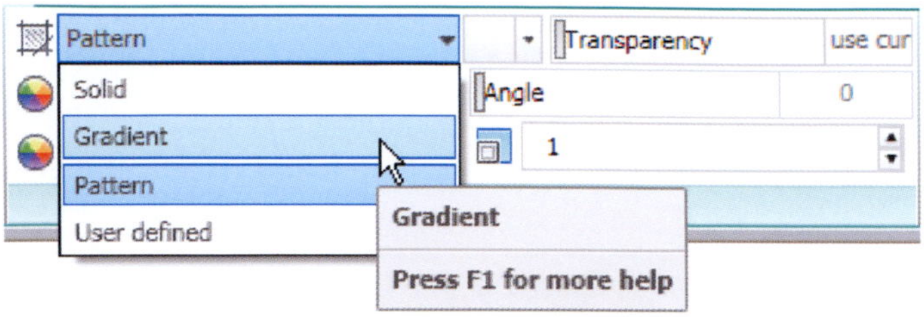

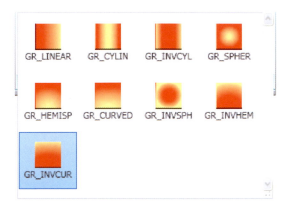

Fig. 8.13 Selecting **Gradient** in the **Hatch Creation/Properties** panel

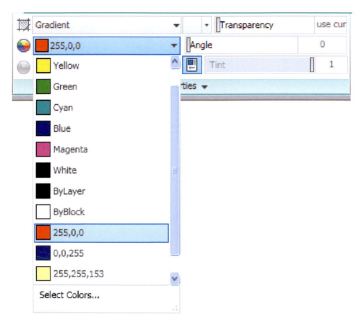

Fig. 8.14 The **Gradient patterns** in the **Hatch Creation/Pattern** panel

4. Change the colour of the **Gradient** patterns with a *click* on the red
   option in the **Select Colors …** drop-down menu in the **Hatch Creation/
   Properties** panel. The hatch patterns all change colour to red (Fig. 8.15).

Fig. 8.15 Changing the colours of the **Gradient** patterns

Fifth example – advanced hatching (Fig. 8.17)

Left-click **Normal Island Detection** in the **Hatch Creation/Options** panel extension. The drop-down shows several forms of **Island** hatching (Fig. 8.16).

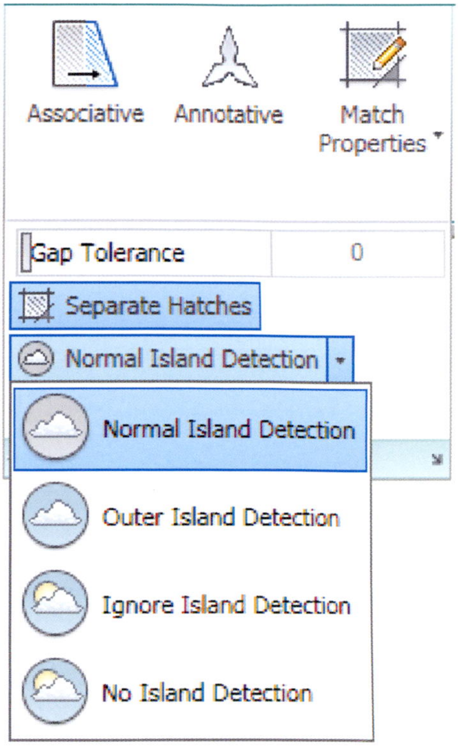

Fig. 8.16 The **Island detection** options in the **Hatch Creation/Options** panel

1. Construct a drawing which includes three outlines as shown in the left-hand drawing of Fig. 8.17 and copy it twice to produce three identical drawings.
2. Select the hatch patterns **STARS** at an angle of **0** and scale **1**.

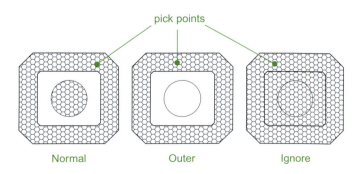

pick points

Normal   Outer   Ignore

Fig. 8.17 Fifth example – advanced hatching

CHAPTER 8

3. *Click* **Normal Island Detection** from the drop-down menu.
4. *Pick* a point in the left-hand drawing. The drawing hatches as shown.
5. Repeat in the centre drawing with **Outer Island Detection** selected.
6. Repeat in the right-hand drawing with **Ignore Island Detection** selected.

Sixth example – text in hatching (Fig. 8.18)

1. Construct a pline rectangle using the sizes given in Fig. 8.18.
2. In the **Text Style Manager** dialog, set the text font to **Arial** and its **Height = 25**.
3. Using the **Dtext** tool *enter* the text as shown central to the rectangle.
4. Hatch the area using the **HONEY** hatch pattern set to an angle of **0** and scale of **1**.

The result is shown in Fig. 8.18.

Fig. 8.18 Sixth example – text in hatching

> **Note**
>
> Text will be entered with a surrounding boundary area free from hatching providing **Normal Island Detection** has been selected from the **Hatch Creation/Options** panel.

REVISION NOTES

1. A large variety of hatch patterns are available when working with AutoCAD 2011.
2. In sectional views in engineering drawings it is usual to show items such as bolts, screws, other cylindrical objects, webs and ribs as outside views.
3. When Associative hatching is set on, if an object is moved within a hatched area, the hatching accommodates to fit around the moved object.
4. Colour gradient hatching is available in AutoCAD 2011.
5. When hatching takes place around text, a space around the text will be free from hatching.

CHAPTER 8

Exercises

Methods of constructing answers to the following exercises can be found in the free website:

http://www.books.elsevier.com/companions/978-0-08-096575-8

1. Fig. 8.19 is a pictorial drawing of the component shown in the orthographic projection Fig. 8.20. Construct the three views but with the font view as a sectional view based on the section plane **A-A**.

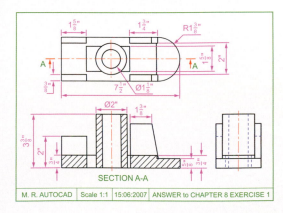

Fig. 8.19 Exercise 1 – a pictorial view

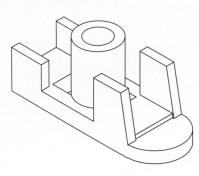

Fig. 8.20 Exercise 1

2. Construct the orthographic projection Fig. 8.21 to the given dimensions with the front view as the sectional view **A-A**.

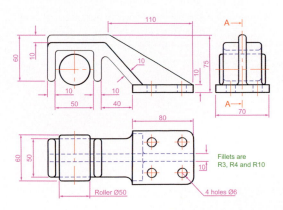

Fig. 8.21 Exercise 2

3. Construct the drawing **Stage 5** following the descriptions of stages given in Fig. 8. 22.

Stage 1
Construct word on Layer 0 and offset on Layer 1

Stage 2
Hatch on Layer HATCH01 with SOLID Turn Layer 0 off

Stage 3
Turn HATCH01 off Turn Layer HATCH02 on Add lines as shown

Stage 4
On HATCH03 Hatch with ANSI31 at Angle 135 and Scale 40 Turn HATCH02 off

Stage 5
Turn HATCH02 on

Fig. 8.22 Exercise 3

CHAPTER 8

4. Fig. 8.23 is a front view of a car with parts hatched. Construct a similar drawing of any make of car, using hatching to emphasise the shape.

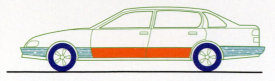

Fig. 8.23 Exercise 4

5. Working to the notes given with the drawing Fig. 8.24, construct the end view of a house as shown. Use your own discretion about sizes for the parts of the drawing.

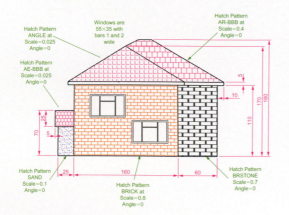

Fig. 8.24 Exercise 5

6. Working to dimensions of your own choice, construct the three-view projection of a two-storey house as shown in Fig. 8.25.

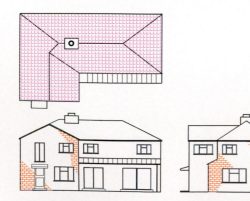

Fig. 8.25 Exercise 6

7. Construct Fig. 8.26 as follows:

a. On layer **Text**, construct a circle of radius **90**.

b. Make layer **0** current.

c. Construct the small drawing to the details as shown and save as a block with a block name **shape** (see Chapter 9).

d. Call the **Divide** tool by *entering* **div** at the command line:

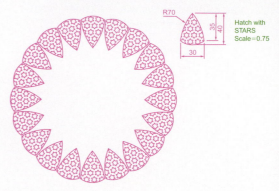

Fig. 8.26 Exercise 7

```
Command: enter div right-click
Select object to divide: pick
   the circle
Enter number of segments or
   [Block]: enter b right-click
Enter name of block to insert:
   enter shape right-click
Align block with object? [Yes/
   No] <Y>: right-click
Enter the number of segments:
   enter 20 right-click
Command
```

e. Turn the layer **Text** off.

Blocks and Inserts

AIMS OF THIS CHAPTER

The aims of this chapter are:

1. To describe the construction of **blocks** and **wblocks** (written blocks).
2. To introduce the insertion of **blocks** and **wblocks** into drawings.
3. To introduce uses of the **DesignCenter** palette.
4. To explain the use of the **Explode** and **Purge** tools.

Introduction

Blocks are drawings which can be inserted into other drawings. Blocks are contained in the data of the drawing in which they have been constructed. Wblocks (written blocks) are saved as drawings in their own right, but can be inserted into other drawings if required.

Blocks

First example – Blocks (Fig. 9.3)

1. Construct the building symbols as shown in Fig. 9.1 to a scale of 1:50.

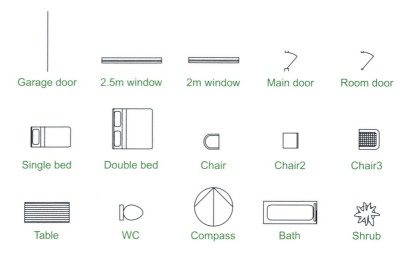

Fig. 9.1 First example – **Blocks** – symbols to be saved as blocks

2. *Left-click* the **Create** tool icon in the **Home/Block** panel (Fig. 9.2).

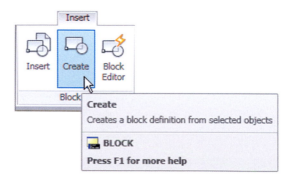

Fig. 9.2 *Click* **Create** tool icon in the **Insert/Block** panel

The **Block Definition** dialog (Fig. 9.3) appears. To make a block from the **Compass** symbol drawing.

a. *Enter* **compass** in the **Name** field.
b. *Click* the **Select Objects** button. The dialog disappears. Window the drawing of the compass. The dialog reappears. Note the icon of the compass at the top-centre of the dialog.
c. *Click* the **Pick Point** button. The dialog disappears. *Click* a point on the compass drawing to determine its **insertion point**. The dialog reappears.
d. If thought necessary *enter* a description in the **Description** field of the dialog.
e. *Click* the **OK** button. The drawing is now saved as a **block** in the drawing.

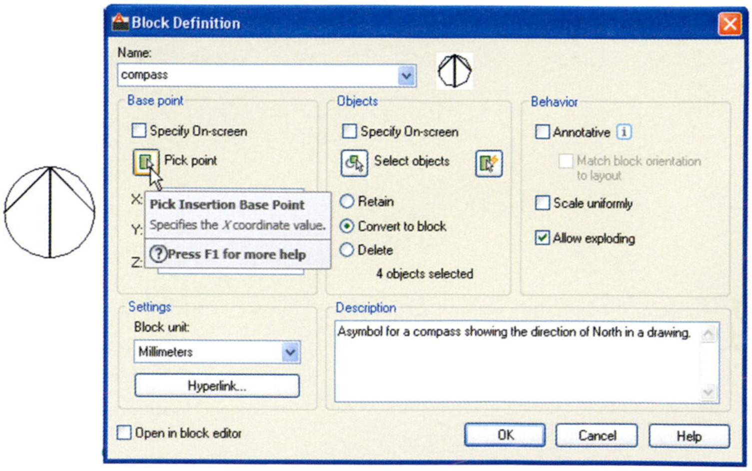

Fig. 9.3 The **Block Definition** dialog with *entries* for the **compass** block

3. Repeat items **1** and **2** to make blocks of all the other symbols in the drawing.
4. Open the **Block Definition** dialog again and *click* the arrow on the right of the **Name** field. Blocks saved in the drawing are listed (Fig. 9.4).

Inserting blocks into a drawing

There are two methods by which symbols saved as blocks can be inserted into another drawing.

Example – first method of inserting blocks

Ensure that all the symbols saved as blocks using the **Create** tool are saved in the data of the drawing in which the symbols were constructed. Erase all

CHAPTER 9

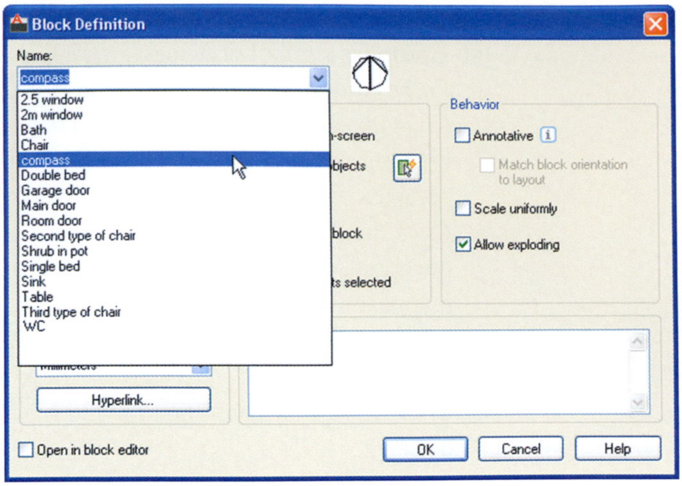

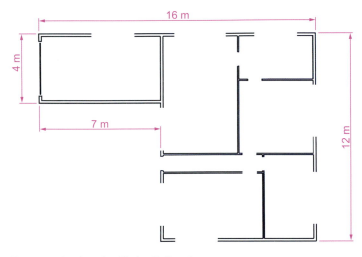

Fig. 9.4 The popup list in the **Name** field of the **Block Definition** dialog

Fig. 9.5 First example – inserting blocks. Outline plan

of the drawings of the symbols and in their place construct the outline of the plan of a bungalow to a scale of 1:50 (Fig. 9.5). Then:

1. *Left-click* the **Insert** tool icon in the **Home/Block** panel (Fig. 9.6) or the **Insert Block** tool in the **Draw** toolbar. The **Insert** dialog appears on screen (Fig. 9.7). From the **Name** popup list select the name of the block which is to be inserted, in this example the **2.5 window**.

2. *Click* the dialog's **OK** button, the dialog disappears. The symbol drawing appears on screen with its insertion point at the intersection of the cursor hairs ready to be *dragged* into its position in the plan drawing.

3. Once all the block drawings are placed, their positions can be adjusted. Blocks are single objects and can thus be dragged into new positions as

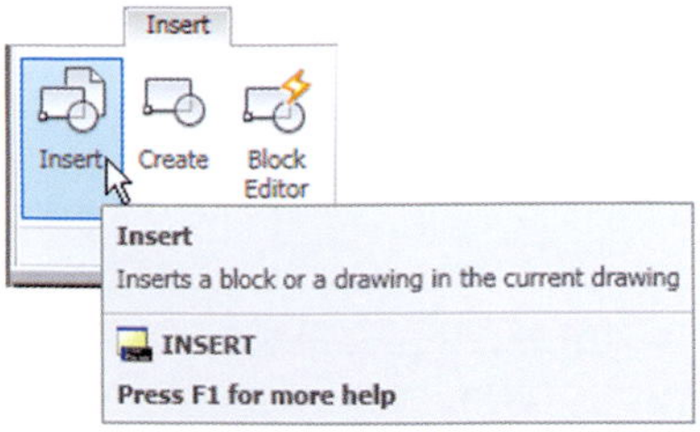

Fig. 9.6 The **Insert** tool icon in the **Home/Block** panel

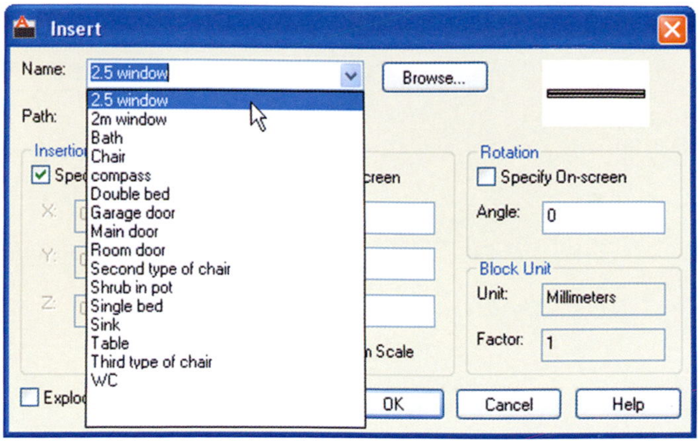

Fig. 9.7 The **Insert** dialog with its **Name** popup list showing all the blocks

required under mouse control. Their angle of position can be amended
at the command line, which shows:

```
Command:_insert
Specify insertion point or [Basepoint/Scale/
  Rotate]: pick
Command:
```

Selection from these prompts allows scaling or rotating as the block is
inserted.

4. Insert all necessary blocks and add other detail as required to the plan
outline drawing. The result is given in Fig. 9.8.

Example – second method of inserting blocks

1. Save the drawing with all the blocks to a suitable file name. Remember
this drawing includes data of the blocks in its file.
2. *Left-click* **DesignCenter** in the **View/Palettes** panel (Fig. 9.9) or press the
Ctrl+2 keys. The **DesignCenter** palette appears on screen (Fig. 9.10).
3. With the outline plan (Fig. 9.5) on screen the symbols can all be
dragged into position from the **DesignCenter**.

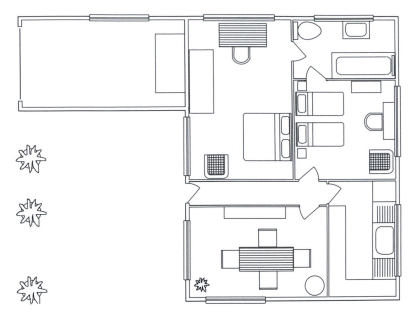

Fig. 9.8 Example – first method of inserting blocks

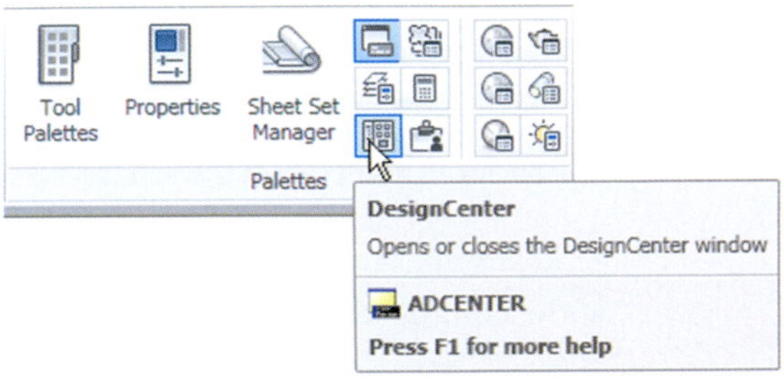

Fig. 9.9 Selecting **DesignCenter** from the **View/Palettes** panel

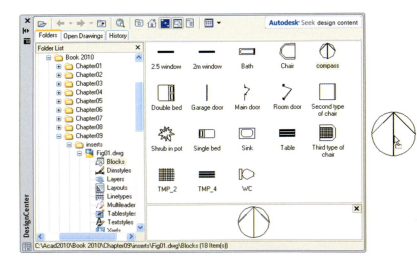

Fig. 9.10 The **DesignCenter** with the compass block *dragged* on screen

Notes about the DesignCenter palette

1. As with other palettes, the **DesignCenter** palette can be resized by *dragging* the palette to a new size from its edges or corners.

2. *Clicks* on one of the three icons at the top-right corner of the palette (Fig. 9.11) have the following results**.

Fig. 9.11 The icons at the top of the **DesignCenter** palette

Tree View Toggle – changes from showing two areas – a **Folder List** and icons of the blocks within a file – to a single area showing the block icons (Fig. 9.12).

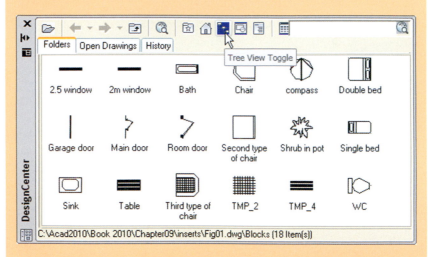

Fig. 9.12 The results of a *click* on **Tree View Toggle**

Preview – a *click* on the icon opens a small area at the base of the palette open showing an enlarged view of the selected block icon.

Description – a *click* on the icon opens another small area with a description of the block.

A block is a single object no matter from how many objects it was originally constructed. This enables a block to be *dragged* about the drawing area as a single object.

CHAPTER 9

The Explode tool

Fig. 9.13 The **Explode** check box in the **Insert** dialog

A check box in the bottom left-hand corner of the **Insert** dialog is labelled **Explode**. If a tick is in the check box, **Explode** will be set on and when a block is inserted it will be exploded into the objects from which it was constructed (Fig. 9.13).

Another way of exploding a block would be to use the **Explode** tool from the **Home/Modify** panel (Fig. 9.14). A *click* on the icon or *entering* **ex** at the command line brings prompts into the command line:

```
Command: _explode
Select objects: <Object Snap Tracking on>  pick a
  block on screen 1 found
Select objects: right-click
Command:
```

And the *picked* object is exploded into its original objects.

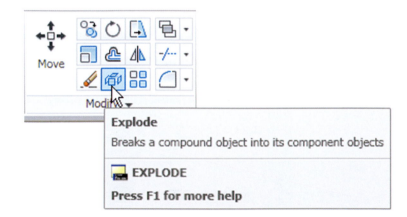

Fig. 9.14 The **Explode** tool icon in the **Home/Modify** panel

Purge

The **Purge** dialog (Fig. 9.15) is called to screen by *entering* **pu** or **purge** at the command line.

Purge can be used to remove data (if any is to be purged) from within a drawing, thus saving file space when a drawing is saved to disk.

To purge a drawing of unwanted data (if any) in the dialog, *click* the **Purge All** button and a sub-dialog appears with three suggestions – purging of a named item, purging of all the items or skip purging a named item.

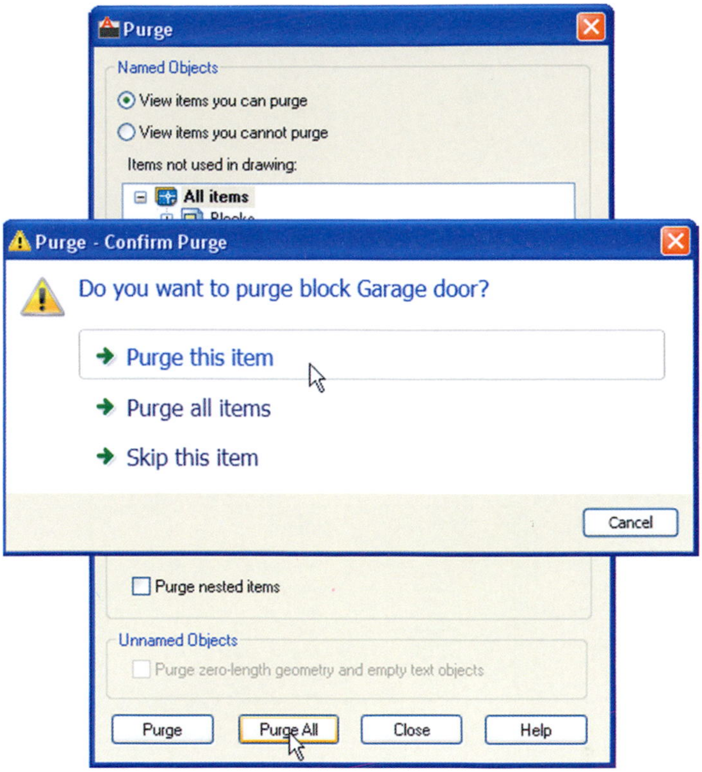

Fig. 9.15 The **Purge** dialog

Taking the drawing Fig. 9.8 as an example. If all the unnecessary data is purged from the drawing, the file will be reduced from **145** Kbytes to **67** Kbytes when the drawing is saved to disk.

Using the DesignCenter (Fig. 9.18)

1. Construct the set of electric/electronic circuit symbols shown in Fig. 9.16 and make a series of blocks from each of the symbols.
2. Save the drawing to a file Fig16.dwg.
3. Open the **acadiso.dwt** template. Open the **DesignCenter** with a *click* on its icon in the **View/Palettes** panel.
4. From the **Folder list** select the file Fig16.dwg and *click* on **Blocks** under its file name. Then *drag* symbol icons from the **DesignCenter** into the drawing area as shown in Fig. 9.17. Ensure they are placed in appropriate positions in relation to each other to form a circuit. If necessary either **Move** or **Rotate** the symbols into correct positions.
5. Close the **DesignCenter** palette with a *click* on the **x** in the top left-hand corner.
6. Complete the circuit drawing as shown in Fig. 9.18.

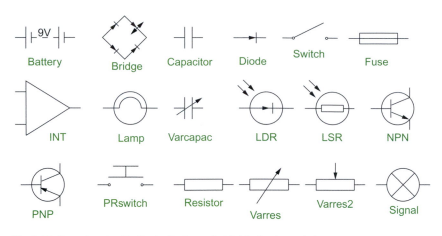

Fig. 9.16 Example using the DesignCenter – electric/electronic symbols

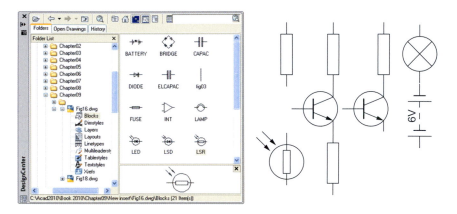

Fig. 9.17 Example using the DesignCenter

Note

Fig. 9.18 does not represent an authentic electronics circuit.

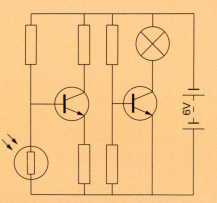

Fig. 9.18 Example using the DesignCenter

Wblocks

Wblocks or written blocks are saved as drawing files in their own right and are not part of the drawing in which they have been saved.

Example – wblock (Fig. 9.19)

1. Construct a light emitting diode (**LED**) symbol and *enter* **w** at the command line. The **Write Block** dialog appears (Fig. 9.19).
2. *Click* the button marked with three full stops (…) to the right of the **File name and path** field and from the **Browse for Drawing File** dialog which comes to screen select an appropriate directory. The directory name appears in the **File name and path** field. Add **LED.dwg** at the end of the name.
3. Make sure the **Insert units** is set to **Millimetres** in its popup list.
4. *Click* the **Select objects** button, Window the symbol drawing and when the dialog reappears, *click* the **Pick point** button, followed by selecting the left-hand end of the symbol.
5. Finally *click* the **OK** button of the dialog and the symbol is saved in its selected directory as a drawing file **LED.dwg** in its own right.

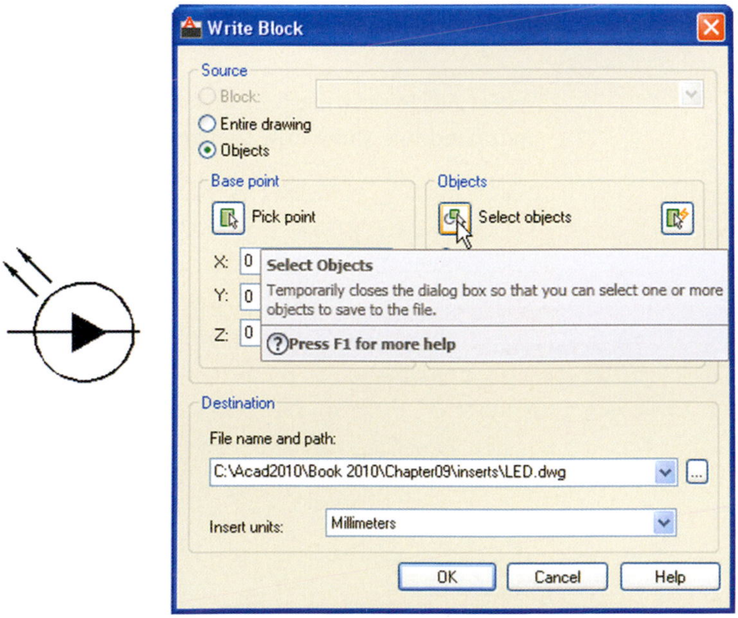

Fig. 9.19 Example – **Wblock**

CHAPTER 9

Note on the DesignCenter

Drawings can be inserted into the AutoCAD window from the **DesignCenter** by dragging the icon representing the drawing into the window (Fig. 9.20).

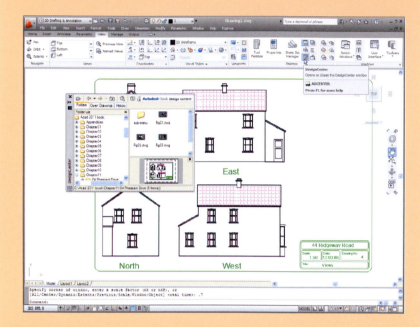

Fig. 9.20 An example of a drawing *dragged* from the **DesignCenter**

When such a drawing is *dragged* into the AutoCAD window, the command line shows a sequence such as:

```
Command: _-INSERT Enter block name or [?]:
 "C:\Acad 2011 book\Chapter11\64
Pheasant Drive\Fig04.dwg"
Units: Millimeters Conversion: 1.0000
Specify insertion point or [Basepoint/Scale/X/
 Y/Z/Rotate]: pick
Enter X scale factor, specify opposite corner,
 or [Corner/XYZ] <1>: right-click
Enter Y scale factor <use X scale factor>:
 right-click
Specify rotation angle <0>: right-click
Command:
```

REVISION NOTES

1. Blocks become part of the drawing file in which they were constructed.
2. Wblocks become drawing files in their own right.
3. Drawings or parts of drawings can be inserted in other drawings with the Insert tool.
4. Inserted blocks or drawings are single objects unless either the Explode check box of the Insert dialog is checked or the block or drawing is exploded with the Explode tool.
5. Drawings can be inserted into the AutoCAD drawing area using the DesignCenter.
6. Blocks within drawings can be inserted into drawings from the DesignCenter.
7. Construct drawings of the electric/electronics symbols in Fig. 9.17 and save them as blocks in a drawing file electronics.dwg.

Exercises

Methods of constructing answers to the following exercises can be found in the free website:

http://books.elsevier.com/companions/978-0-08-096575-8

1. Construct the building symbols (Fig. 9.21) in a drawing saved as **symbols.dwg**. Then using the **DesignCenter** construct a building drawing of the first floor of the house you are living in making use of the symbols. Do not bother too much about dimensions because this exercise is designed to practise using the idea of making blocks and using the **DesignCenter**.

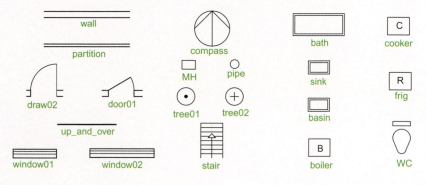

Fig. 9.21 Exercise 1

2. Construct drawings of the electric/electronics symbols in Fig. 9.17 (page 186) and save them in a drawing file **electronics.dwg**.

3. Construct the electronics circuit given in Fig. 9.22 from the file **electronics.dwg** using the **DesignCenter**.

4. Construct the electronics circuit given in Fig. 9.23 from the file **electronics.dwg** using the **DesignCenter**.

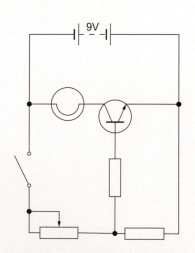

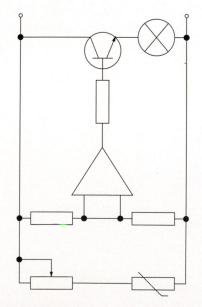

Fig. 9.22 Exercise 3

Fig. 9.23 Exercise 4

Chapter 10

Other types of file format

AIMS OF THIS CHAPTER

The aims of this chapter are:

1. To introduce Object Linking and Embedding (**OLE**) and its uses.
2. To introduce the use of Encapsulated Postscript (**EPS**) files.
3. To introduce the use of Data Exchange Format (**DXF**) files.
4. To introduce raster files.
5. To introduce **Xrefs**.

Object Linking and Embedding

First example – Copying and Pasting (Fig. 10.3)

1. Open any drawing in the AutoCAD 2011 window (Fig. 10.1).

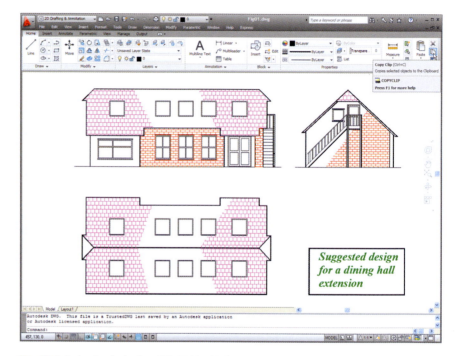

Fig. 10.1 A drawing in the AutoCAD 2011 with **Copy Clip** selected

2. *Click* **Copy Clip** from the **Home/Clipboard** panel. The command line shows:

```
Command: _copyclip
Select objects: left-click top left of the
  drawing
Specify opposite corner: left-click bottom right
  of the drawing 457 found
Select objects: right-click
Command:
```

3. Open **Microsoft Word** and *click* on **Paste** in the **Edit** drop-down menu (Fig. 10.2). The drawing from the **Clipboard** appears in the **Microsoft Word** document. Add text as required.

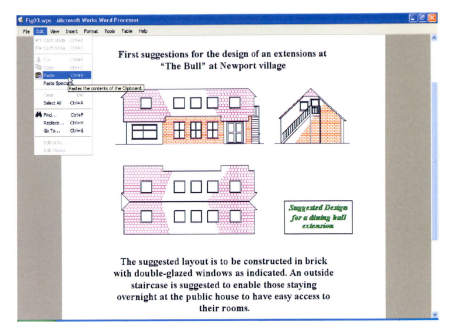

Fig. 10.2 Example – **Copying and Pasting**

> **Note**
>
> Similar results can be obtained using the **Copy**, **Copy Link** or **Copy with Base Point** tools from the **Edit** drop-down menu.

Second example – EPS file (Fig. 10.5)

1. With the same drawing on screen *click* on **Export…** in the **File** drop-down menu (Fig. 10.3) or *click* **Export/Other formats** in the menu appearing with a *click* on the **A** icon at the top left-hand corner of the AutoCAD window. The **Export Data** dialog appears (Fig. 10.3). *Pick* **Encapsulated PS (*.eps)** from the **Files of type** popup list then *enter* a suitable file name (e.g. **building.eps**) in the **File name field** and *click* the **Save** button.

2. Open a desktop publishing application. That shown in Fig. 10.4 is **PageMaker**.

3. From the **File** drop-down menu of **PageMaker** *click* **Place…** A dialog appears listing files which can be placed in a PageMaker document. Among the files named will be **building.eps**. *Double-click* that file

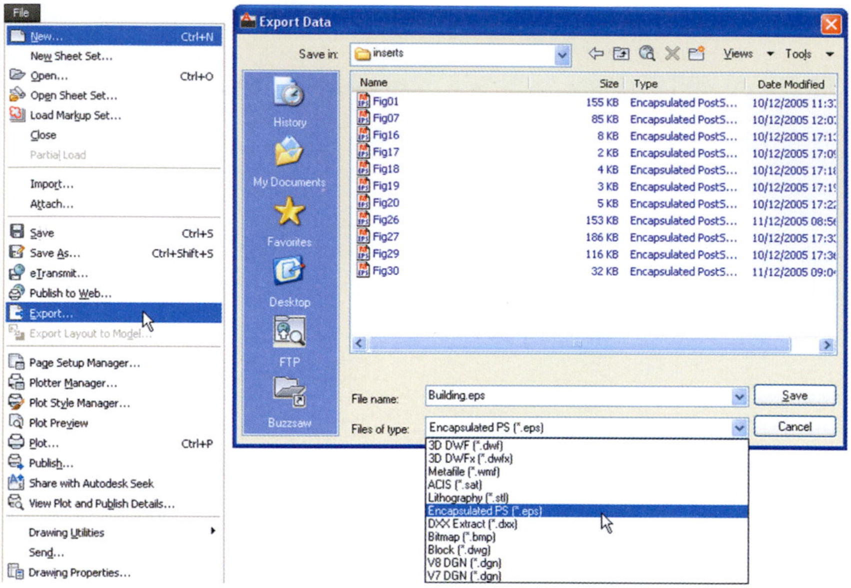

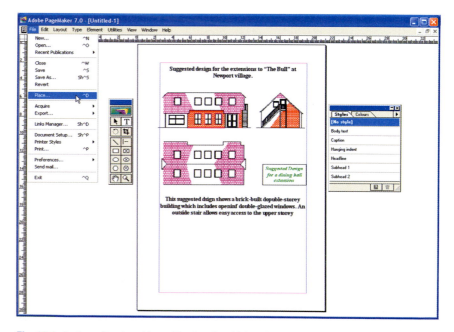

Fig. 10.3 The **Export** tool icon from the **File** drop-down menu and the **Export Data** dialog

name and an icon appears the placing of which determines the position of the ***eps** file drawing in the PageMaker document (Fig. 10.4).

4. Add text as required.
5. Save the PageMaker document to a suitable file name.
6. Go back to the AutoCAD drawing and delete the title.

Fig. 10.4 An ***eps** file placed in position in a PageMaker document

7. Make a new ***.eps** file with the same file name (**building.eps**).

8. Go back into **PageMaker** and *click* **Links Manager…** in the **File** drop-down menu. The **Links Manager** dialog appears (Fig. 10.5). Against the name of the **building.eps** file name is a dash and a note at the bottom of the dialog explaining that changes have taken place in the drawing from which the ***eps** had been derived. *Click the* **Update** button and when the document reappears the drawing in PageMaker no longer includes the erased title.

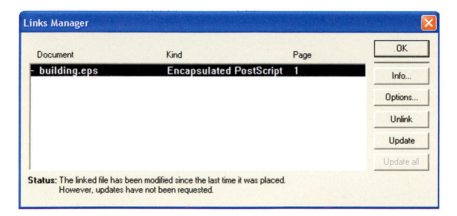

Fig. 10.5 The **Links Manager** dialog of PageMaker

Notes

1. This is **Object Linking and Embedding** (**OLE**). Changes in the AutoCAD drawing saved as an ***eps** file are linked to the drawing embedded in another application document, so changes made in the AutoCAD drawing are reflected in the PageMaker document.

2. There is actually no need to use the **Links Manager** because if the file from PageMaker is saved with the old ***eps** file in place, when it is reopened the file will have changed to the redrawn AutoCAD drawing, without the erased title.

DXF (data exchange format) files

The ***.DXF** format was originated by Autodesk (publishers of AutoCAD), but is now in general use in most **CAD** (Computer Aided Design) software.

A drawing saved to a *.dxf format file can be opened in most other CAD software applications. This file format is of great value when drawings are being exchanged between operators using different CAD applications.

Example – DXF file (Fig. 10.7)

1. Open a drawing in AutoCAD. This example is shown in Fig. 10.6.

Fig. 10.6 Example – **DXF file**. Drawing to be saved as a dxf file

2. *Click* on **Save As…** in the **Menu Browser** dialog and in the **Save Drawing As** dialog which appears, *click* **AutoCAD 2010 DXF [*.dxf]** in the **Files of type** field popup list.
3. *Enter* a suitable file name. In this example this is **Fig06.dxf**. The extension **.dxf** is automatically included when the **Save** button of the dialog is *clicked* (Fig. 10.7).
4. The **DXF** file can now be opened in the majority of CAD applications and then saved to the drawing file format of the CAD in use.

Note

To open a **DXF** file in AutoCAD 2011, select **Open…** from the **Menu Browser** dialog and in the **Select File** dialog select **DXF [*.dxf]** from the popup list from the **Files of type** field.

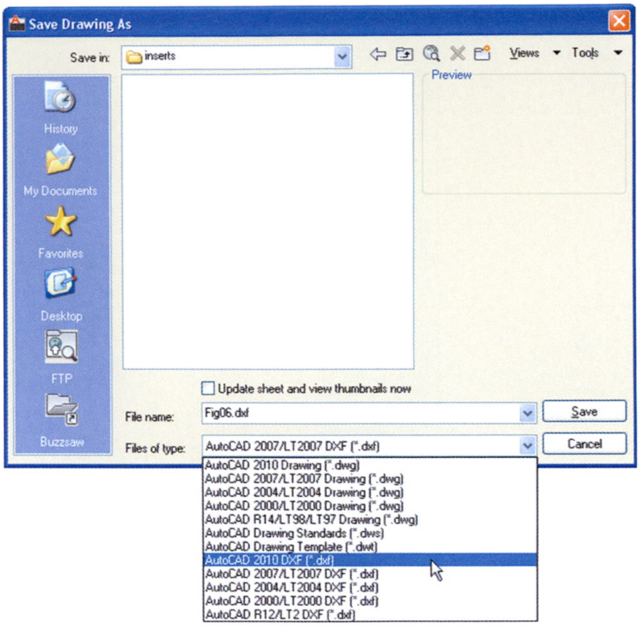

Fig. 10.7 The **Save Drawing As** dialog set to save drawings in **DXF** format

Raster images

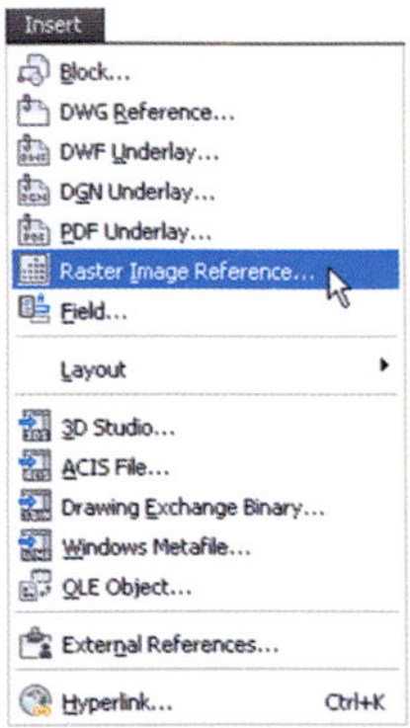

Fig. 10.8 Selecting **Raster Image Reference...** from the **Insert** drop-down menu

A variety of raster files can be placed into AutoCAD 2011 drawings from the **Select Image File** dialog brought to screen with a *click* on **Raster Image Reference...** from the **Insert** drop-down menu. In this example the selected raster file is a bitmap (extension ***.bmp**) of a rendered 3D model drawing.

Example – placing a raster file in a drawing (Fig. 10.11)

1. *Click* **Raster Image Reference...** from the **Insert** drop-down menu (Fig. 10.8). The **Select Reference File** dialog appears (Fig. 10.9). *Click* the file name of the image to be inserted, **Fig05** (a bitmap *.bmp). A preview of the bitmap appears.
2. *Click* the **Open** button of the dialog. The **Attach Image** dialog appears (Fig. 10.10) showing a preview of the bitmap image.
3. *Click* the **OK** button, the command line shows:

CHAPTER 10

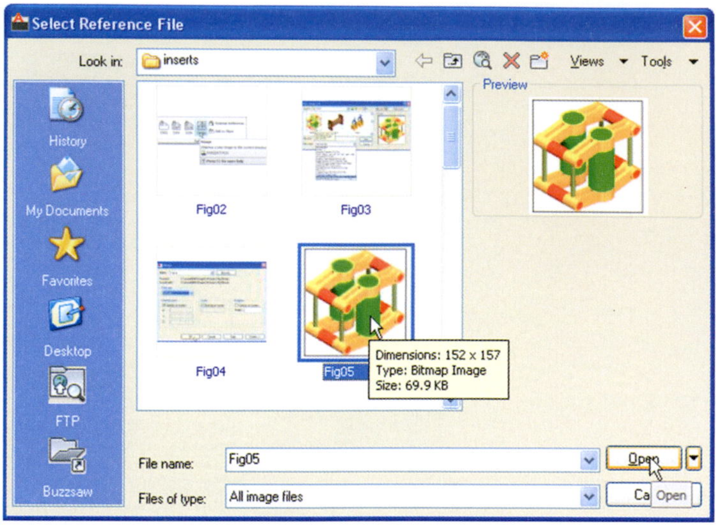

Fig. 10.9 The **Select Reference File** dialog

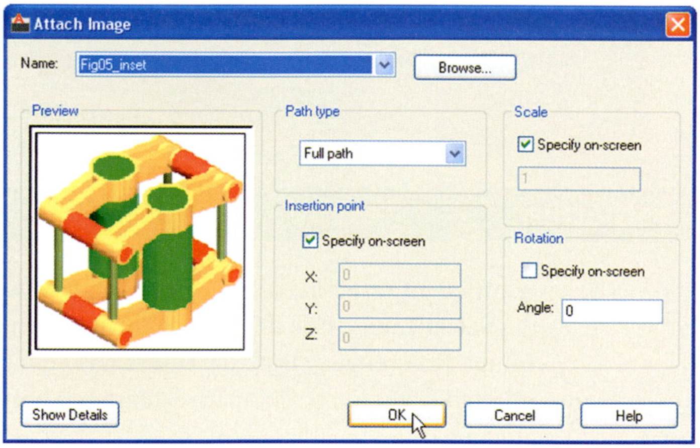

Fig. 10.10 The **Attach Image** dialog

```
Command: _imageattach
Specify insertion point <0,0>: click at a point on
  screen
Base image size: Width: 1.000000, Height:
  1.032895, Millimetres
Specify scale factor <1>: drag a corner of the
  image to obtain its required size
Command:
```

And the raster image appears at the *picked* point (Fig. 10.11).

CHAPTER 10

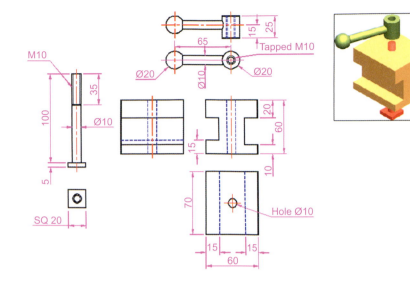

Fig. 10.11 Example – placing a raster file in a drawing

Notes

As will be seen from the **Insert** drop-down menu and the dialogs which can be opened from the menu, a variety of different types of images can be inserted into an AutoCAD drawing. Some examples are:

External References (Xrefs) – If a drawing is inserted into another drawing as an external reference, any changes made in the original xref drawing are automatically reflected in the drawing into which the xref has been inserted. See later in this chapter.

Field – A *click* on the name brings up the **Field** dialog. Practise inserting various categories of field names from the dialog.

Layout – A wizard appears allowing new layouts to be created and saved for new templates if required.

3D Studio – allows the insertion of images constructed in the Autodesk software **3D Studio** from files with the format ***.3ds**.

External references (Xrefs)

If a drawing is inserted into another drawing as an external reference, any changes made in the original Xref drawing subsequent to its being inserted are automatically reflected in the drawing into which the Xref has been inserted.

CHAPTER 10

Example – External References (Fig. 10.19)

1. Construct the three-view orthographic drawing Fig. 10.12. Dimensions for this drawing will be found in Fig. 15.52. Save the drawing to a suitable file name.

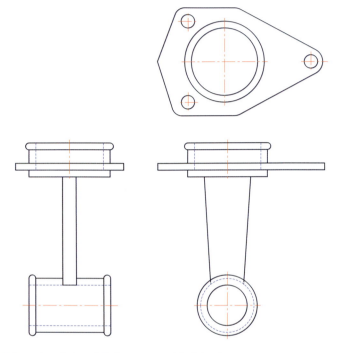

Fig. 10.12 Example – **External References** – original drawing

Fig. 10.13 The spindle drawing saved as **Fig13.dwg**

2. As a separate drawing construct Fig. 10.13. Save it as a wblock with the name of **Fig13.dwg** and with a base insertion point at the crossing of its centre line with the left-hand end of its spindle.

3. Click **External References** in the **View/Palettes** panel (Fig. 10.14). The **External Reference** palette appears (Fig. 10.15).

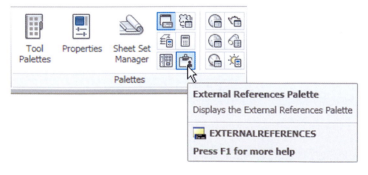

Fig. 10.14 The **External Reference** tool in the **View/Palettes** panel

Fig. 10.15 The **External References** palette

4. *Click* its **Attach** button and select **Attach DWG...** from the popup list which appears when a *left-click* is held on the button. Select the drawing of a spindle (**Fig13.dwg**) from the **Select Reference file** dialog which appears followed by a *click* on the dialog's **Open** button. This brings up the **Attach External Reference** dialog (Fig. 10.16) showing **Fig13** in its **Name** field. *Click* the dialog's **OK** button.

5. The spindle drawing appears on screen ready to be *dragged* into position. Place it in position as indicated in Fig. 10.17.

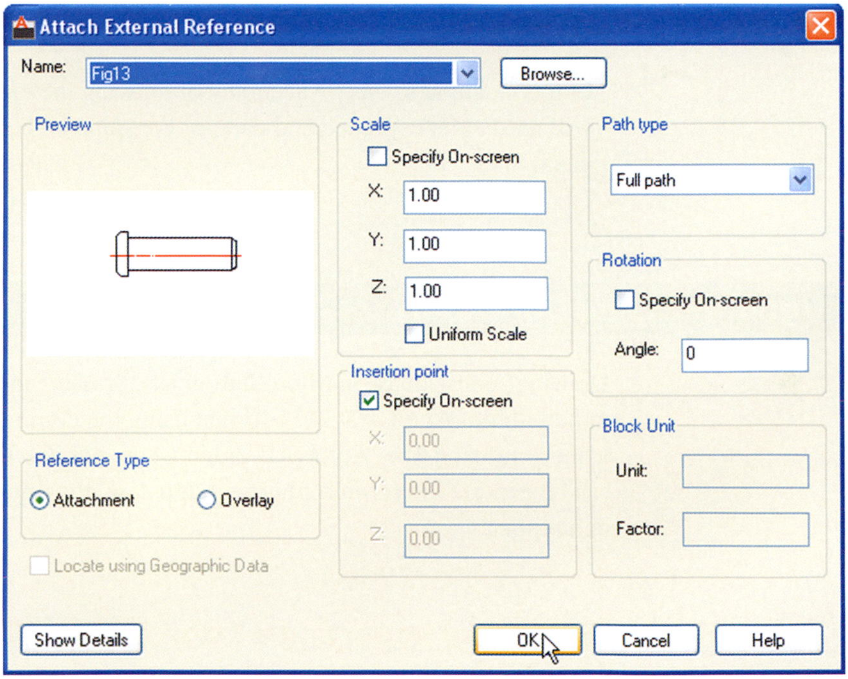

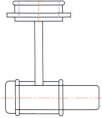

Fig. 10.17 The spindle in place in the original drawing

Fig. 10.16 The **Attach External Reference** dialog

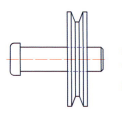

Fig. 10.18 The revised **spindle.dwg** drawing

6. Save the drawing with its xref to its original file name.
7. Open **Fig15.dwg** and make changes as shown in Fig. 10.18.
8. Now reopen the original drawing. The **external reference** within the drawing has changed in accordance with the alterations to the spindle drawing. Fig. 10.19 shows the changes in the front view of the original drawing.

CHAPTER 10

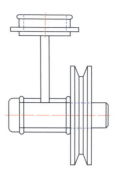

Fig. 10.19 Example – **Xrefs**

Note

In this example to ensure accuracy of drawing the **external reference** will need to be exploded and parts of the spindle changed to hidden detail lines.

Dgnimport and Dgnexport

Drawings constructed in MicroStation V8 format (***.dgn**) can be imported into AutoCAD 2011 format using the command **dgnimport** at the command line. AutoCAD drawings in AutoCAD 2004 format can be exported into MicroStation ***.dgn** format using the command **dgnexport**.

Example of importing a *.dgn drawing into AutoCAD

1. Fig. 10.20 is an example of an orthographic drawing constructed in MicroStation V8.
2. In AutoCAD 2011 at the command line *enter* **dgnimport**. The dialog Fig. 10.21 appears on screen from which the required drawing file name can be selected. When the **Open** button of the dialog is *clicked* a warning window appears informing the operator of steps to take in order to load the drawing. When completed the drawing loads Fig. 10.22).

In a similar manner AutoCAD drawing files can be exported to MicroStation using the command **dgnexport** *entered* at the command line.

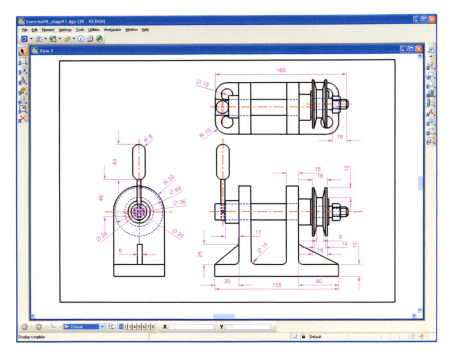

Fig. 10.20 Example – a drawing in MicroStation V8

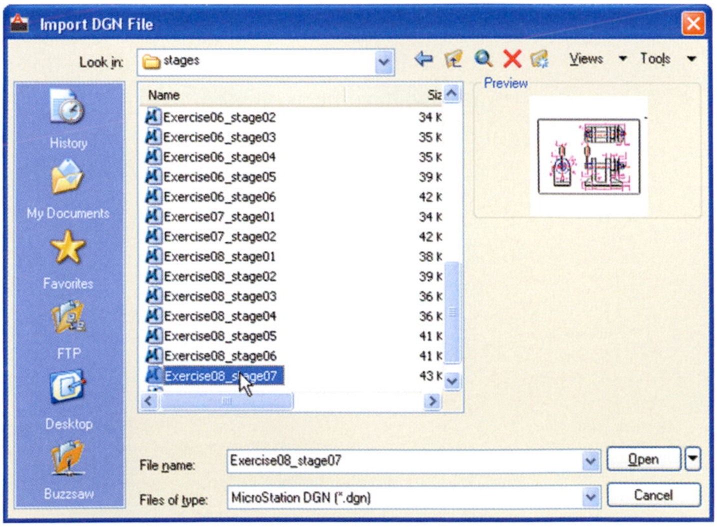

Fig. 10.21 The **Import DGN File** dialog

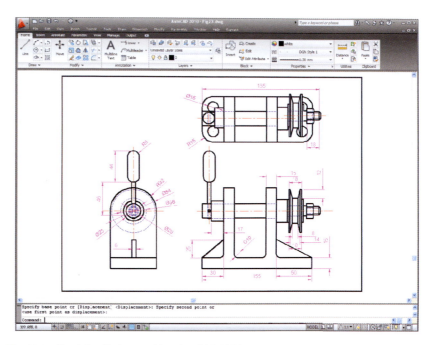

Fig. 10.22 The ***.dgn** file imported into AutoCAD 2011

Multiple Design Environment

1. Open several drawings in AutoCAD, in this example four separate drawings have been opened.
2. In the **View/Windows** panel *click* **Tile Horizontally** (Fig. 10.23). The four drawings rearrange as shown in Fig. 10.24.

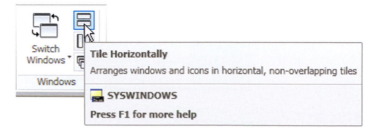

Fig. 10.23 Selecting **Tile Horizontally** from the **View/Windows** panel

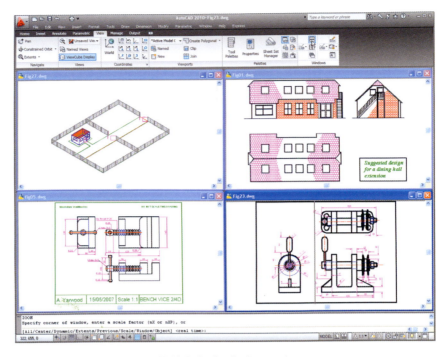

Fig. 10.24 Four drawings in the **Multiple Design Environment**

REVISION NOTES

1. The Edit tools Copy Clip, Copy with Base Point and Copy Link to enable objects from AutoCAD 2011 to be copied for Pasting onto other applications.
2. Objects can be copied from other applications to be pasted into the AutoCAD 2011 window.
3. Drawings saved in AutoCAD as DXF (*.dxf) files can be opened in other Computer Aided Design (CAD) applications.
4. Similarly drawings saved in other CAD applications as *.dxf files can be opened in AutoCAD 2011.
5. Raster files of the format types *.bmp, *.jpg, *pcx, *.tga, *.tif among other raster type file objects can be inserted into AutoCAD 2011 drawings.
6. Drawings saved to the Encapsulated Postscript (*.eps) file format can be inserted into documents of other applications.
7. Changes made in a drawing saved as an *.eps file will be reflected in the drawing inserted as an *.eps file in another application.
8. When a drawing is inserted into another drawing as an external reference, changes made to the inserted drawing will be updated in the drawing into which it has been inserted.
9. A number of drawings can be opened at the same time in the AutoCAD 2011 window.
10. Drawings constructed in MicroStation V8 can be imported into AutoCAD 2011 using the command dgnimport.
11. Drawings constructed in AutoCAD 2011 can be saved as MicroStation *.dgn drawings to be opened in MicroStation V8.

Exercises

Methods of constructing answers to the following exercises can be found in the free website:

http://books.elsevier.com/companions/978-0-08-096575-8

1. Fig. 10.25 shows a pattern formed by inserting an external reference and then copying or arraying the external reference.

Fig. 10.25 Exercise 1 – original pattern

The hatched parts of the external reference drawing were then changed using a different hatch pattern. The result of the change in the hatching is shown in Fig. 10.26.

Fig. 10.26 Exercise 1

Construct a similar xref drawing, insert as an xref, array or copy to form the pattern, then change the hatching, save the xref drawing and note the results.

2. Fig. 10.27 is a rendering of a roller between two end holders. Fig. 10.28 gives details of the end holders and the roller in orthographic projections.

Fig. 10.27 Exercise 2 – a rendering of the holders and roller

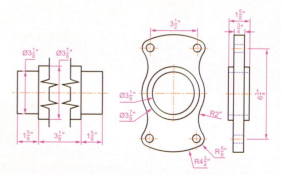

Fig. 10.28 Exercise 2 – details of the parts of the holders and roller

Construct a full size front view of the roller and save to a file name roller.dwg. Then as a separate drawing construct a front view of the two end holders in their correct positions to receive the roller and save to the file name assembly.dwg.

Insert the roller drawing into the assembly drawing as an xref.

Open the roller.dwg and change its outline as shown in Fig. 10.29. Save the drawing. Open the assembly.dwg and note the change in the inserted xref.

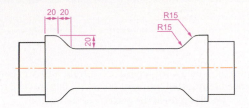

Fig. 10.29 The amended Xref drawing

3.   Click Image… in the Reference panel and insert a JPEG image (*.jpg file) of a photograph into the AutoCAD 2010 window. An example is given in Fig. 10.30.

Fig. 10.30 Exercise 3 – an example

4. Using Copy from the Insert drop-down menu, copy a drawing from AutoCAD 2010 into a Microsoft Word document. An example is given in Fig. 10.31. Add some appropriate text.

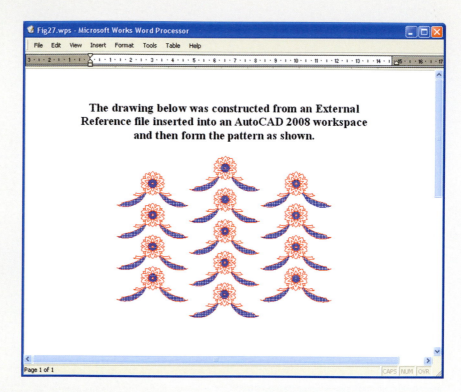

Fig. 10.31 Exercise 4 – an example

5. The plan in Figs 10.1–10.3 is incorrect in that some details have been missed from the drawing. Can you identify the error?

Sheet sets

The aims of this chapter are:

1. To introduce sheet sets.
2. To describe the use of the **Sheet Set Manager**.
3. To give an example of a sheet set based on the design of a two-storey house.

Sheet sets

When anything is to be manufactured or constructed, whether it be a building, an engineering design, an electronics device or any other form of manufactured artefact, a variety of documents, many in the form of technical drawings, will be needed to convey to those responsible for constructing the design and all the information necessary to be able to proceed according to the wishes of the designer. Such sets of drawings may be passed between the people or companies responsible for the construction, enabling all those involved to make adjustments or suggest changes to the design. In some cases there may well be a considerable number of drawings required in such sets of drawings. In AutoCAD 2011 all the drawings from which a design is to be manufactured can be gathered together in a **sheet set**. This chapter shows how a much reduced sheet set of drawings for the construction of a house at 62 Pheasant Drive can be produced. Some other drawings, particularly detail drawings, would be required in this example, but to save page space, the sheet set described here consists of only four drawings with a subset of another four drawings.

Sheet set for 62 Pheasant Drive

1. Construct a template **62 Pheasant Drive.dwt** based upon the acadiso. dwt template, but including a border and a title block. Save the template in a **Layout1** format. An example of the title block from one of the drawings constructed in this template is shown in Fig. 11.1.

Fig. 11.1 The title block from Drawing number **2** of the sheet set drawings

2. Construct each of the drawings which will form the sheet set in this drawing template. The whole set of drawings is shown in Fig. 11.2. Save the drawings in a directory – in this example this has been given the name **62 Pheasant Drive**.
3. *Click* **Sheet Set Manager** in the **View/Palettes** panel (Fig. 11.3). The **Sheet Set Manager** palette appears (Fig. 11.4). *Click* **New Sheet Set…** in the popup menu at the top of the palettes. The first of a series of **Create Sheet Set** dialogs appears – the **Create Sheet Set – Begin**

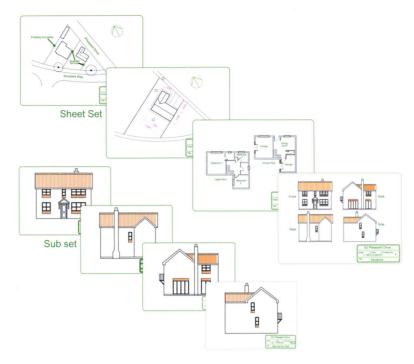

Fig. 11.2 The eight drawings in the **62 Pheasant Drive** sheet set

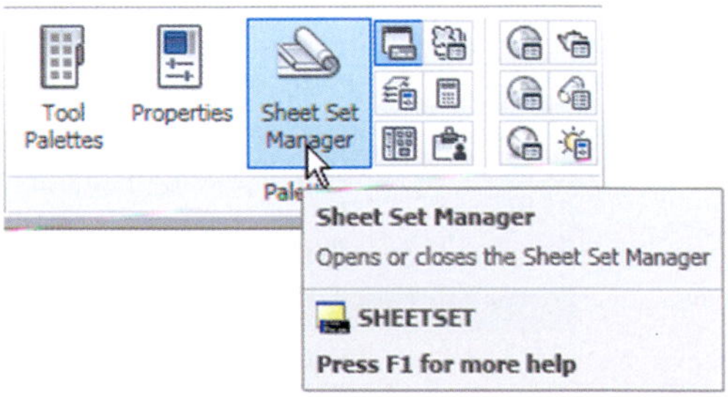

Fig. 11.3 Selecting **Sheet Set Manager** from the **View/Palettes** panel

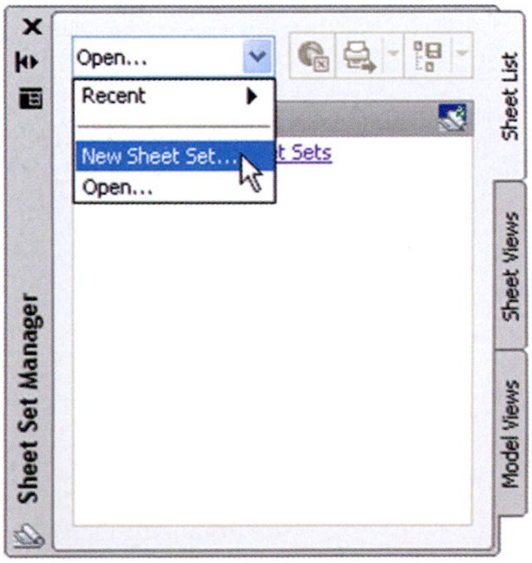

Fig. 11.4 The **Sheet Set Manager** palette

CHAPTER 11

dialog (Fig. 11.5). *Click* the radio button next to **Existing drawings**, followed by a *click* on the **Next** button and the next dialog **Sheet Set Details** appears (Fig. 11.6).

4. *Enter* details as shown in the dialog as shown in Fig. 11.6. Then *click* the **Next** button to bring the **Choose Layouts** dialog to screen (Fig. 11.7).

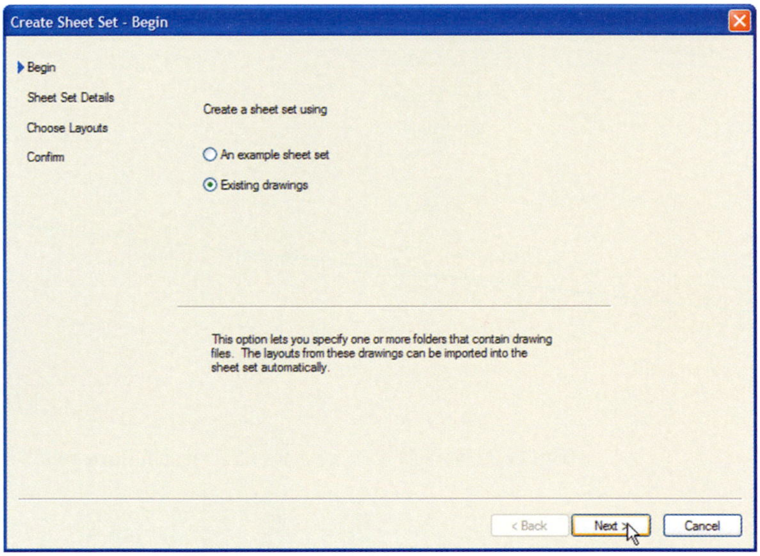

Fig. 11.5 The first of the **Create Sheet Set** dialogs – **Begin**

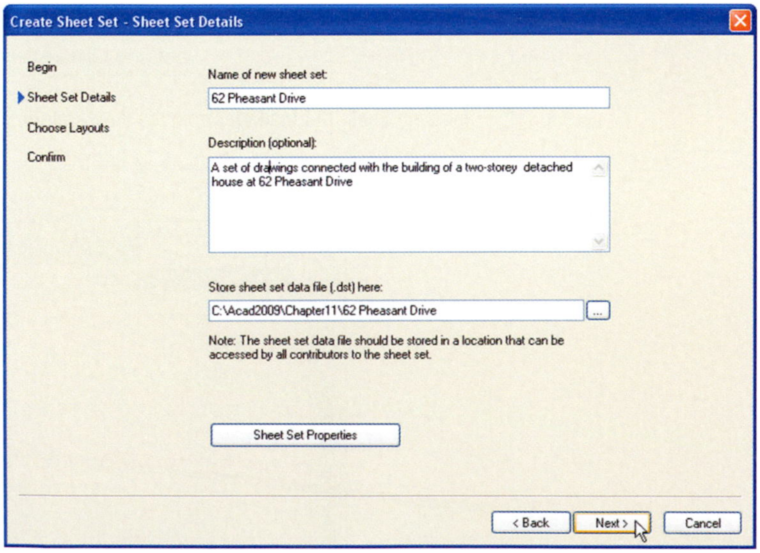

Fig. 11.6 The **Sheet Set Details** dialog

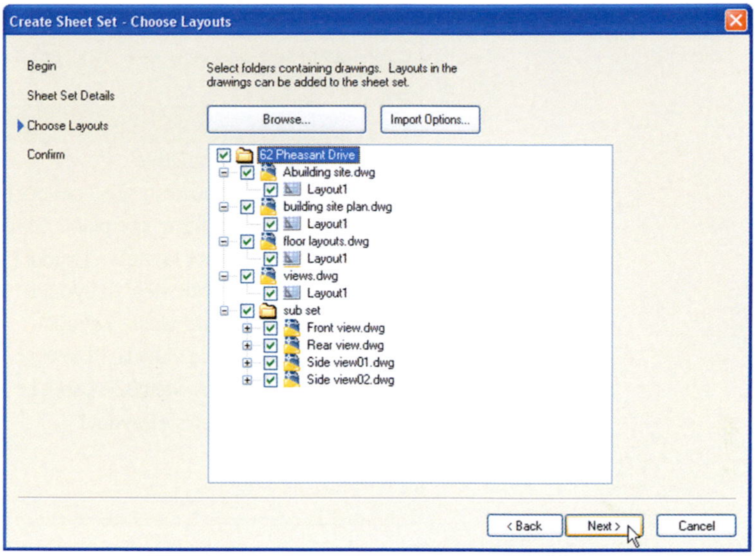

Fig. 11.7 The **Choose Layouts** dialog

5. *Click* its **Browse** button and from the **Browse for Folder** list which comes to screen, *pick* the directory **62 Pheasant Drive**. *Click* the **OK** button and the drawings held in the directory appears in the **Choose Layouts** dialog (Fig. 11.7). If satisfied the list is correct, *click* the **Next** button. A **Confirm** dialog appears (Fig. 11.8). If satisfied *click* the **Finish** button and the **Sheet Set Manager** palette appears showing the drawings which will be in the **62 Pheasant Drive** sheet set (Fig. 11.9).

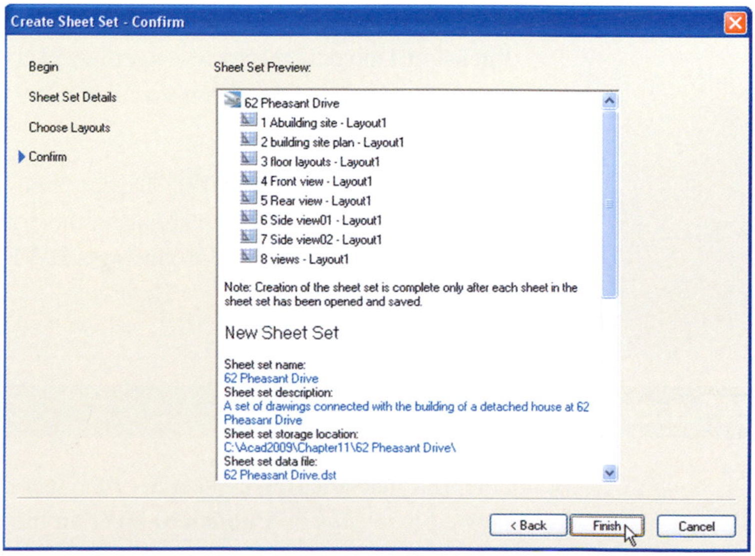

Fig. 11.8 The **Confirm** dialog

CHAPTER 11

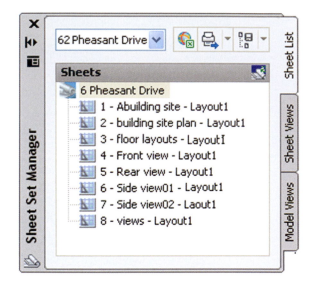

Fig. 11.9 The **Sheet Manager** palette for **62 Pheasant Drive**

Notes

1. The eight drawings in the sheet set are shown in Fig. 11.9. If any of the drawings in the sheet set are subsequently amended or changed, when the drawings is opened again from the **62 Pheasant Drive Sheet Manager** palette, the drawing will include any changes or amendments.

2. Drawings can only be placed into sheet sets if they have been saved in a **Layout** screen. Note that all the drawings shown in the **62 Pheasant Drive** Sheet Set Manager have **Layout1** after the drawing names because each has been saved after being constructed in a **Layout1** template.

3. Sheet sets in the form of **DWF** (Design Web Format) files can be sent via email to others who are using the drawings or placed on an intranet. The method of producing a **DWF** for the **62 Pheasant Drive** Sheet Set follows.

62 Pheasant Drive DWF

1. In the **62 Pheasant Drive** Sheet Set Manager *click* the **Publish** icon, followed by a *click* on **Publish to DWF** in the menu which appears (Fig. 11.10). The **Specify DWF File** dialog appears (Fig. 11.11). *Enter* **62 Pheasant Drive** in the **File name** field followed by a *click*

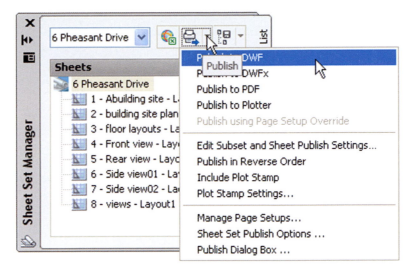

Fig. 11.10 The **Publish** icon in the **Sheet Set Manager**

Fig. 11.11 The **Select DWF File** dialog

on the **Select** button. A warning window (Fig. 11.12) appears. *Click* its **Close** button. The **Publish Job in Progress** icon in the bottom right-hand corner of the AutoCAD 2011 window starts fluctuating in shape showing that the DWF file is being processed (Fig. 11.12). When the icon becomes stationary *right-click* the icon and *click* **View Plotted File...** in the right-click menu which appears (Fig. 11.13).

CHAPTER 11

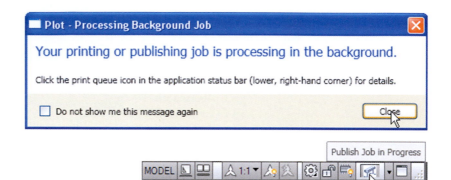

Fig. 11.12 The **Publish Job in Progress** icon

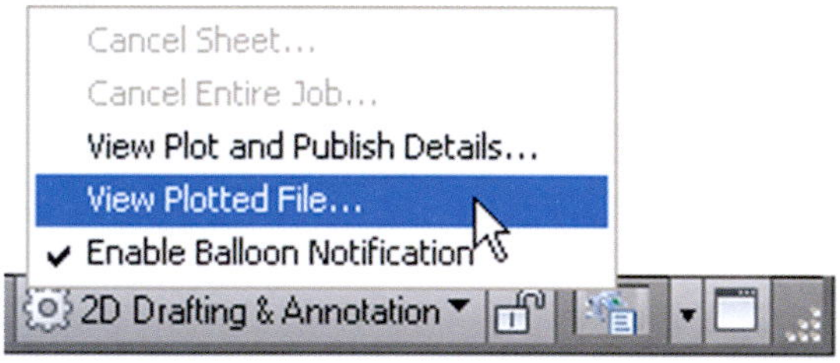

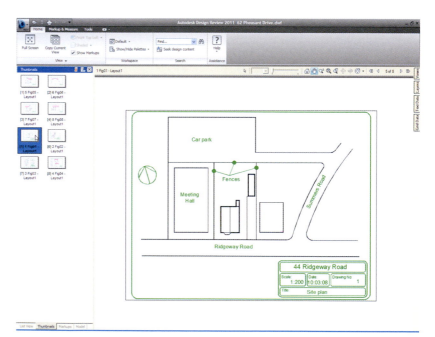

Fig. 11.13 The *right-click* menu of the icon

2. The **Autodesk Design Review** window appears showing the
    62 Pheasant Drive.dwf file (Fig. 11.14). *Click* on the arrow **Next Page**
    (**Page on**) to see other drawings in the DWF file.

Fig. 11.14 The **Autodesk Design Review** showing details of the **62 Pheasant Drive.dwf** file

3. If required the Design Review file can be sent between people by email as an attachment, opened in a company's intranet or, indeed, included within an internet web page.

REVISION NOTES

1. To start off a new sheet set, select the Sheet Set Manager icon in the Tools/Palettes panel.
2. Sheet sets can only contain drawings saved in Layout format.
3. Sheet sets can be published as Design Review Format (*.dwf) files which can be sent between offices by email, published on an intranet or published on a web page.
4. Subsets can be included in sheet sets.
5. Changes or amendments made to any drawings in a sheet set are reflected in the sheet set drawings when the sheet set is opened.

Exercises

Methods of constructing answers to the following exercises can be found in the free website:

http://books.elsevier.com/companions/978-0-08-096575-8

1. Fig. 11.15 is an exploded orthographic projection of the parts of a piston and its connecting rod. There are four parts in the assembly. Small drawings of the required sheet set are shown in Fig. 11.17.

    Construct the drawing Fig. 11.15 and also the four drawings of its parts. Save each of the drawings in a **Layout1** format and construct the sheet set which contains the five drawings (Fig. 11.17).

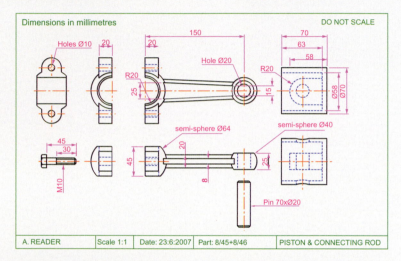

Fig. 11.15 Exercise 1 – exploded orthographic projection

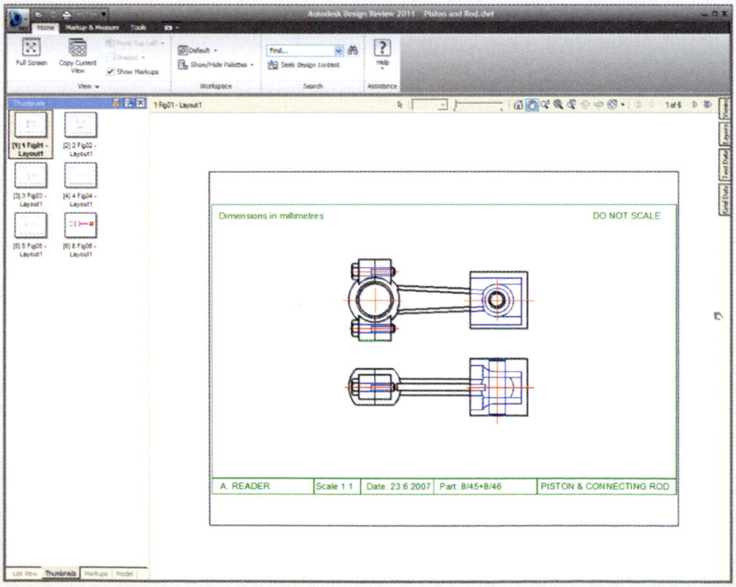

Fig. 11.16 The **DWF** for exercise 1

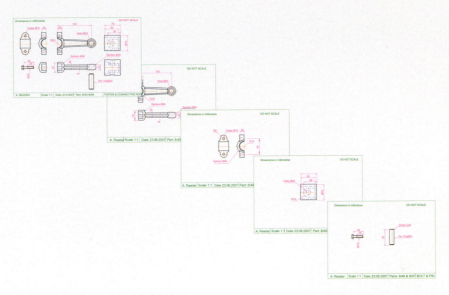

Fig. 11.17 Exercise 1 – the five drawings in the sheet set

Construct the **DWF** file of the sheet set. Experiment sending it to a friend via email as an attachment to a document, asking him/her to return the whole email to you without changes. When the email is returned, open its DWF file and *click* each drawing icon in turn to check the contents of the drawings.

2. Construct a similar sheet set as in the answer to Exercise 1 from the exploded orthographic drawing of a **Machine adjusting spindle** given in Fig. 11.18.

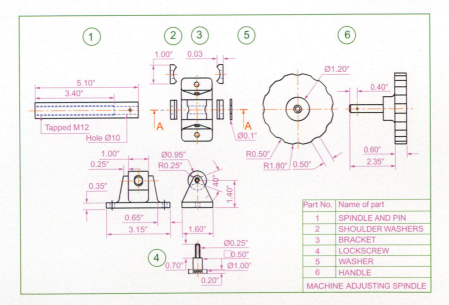

Fig. 11.18 Exercise 2

Part 2

3D Design

Introducing 3D modeling

AIMS OF THIS CHAPTER

The aims of this chapter are:

1. To introduce the tools used for the construction of 3D solid models.
2. To give examples of the construction of 3D solid models using tools from the **Home/Create** panel.
3. To give examples of 2D outlines suitable as a basis for the construction of 3D solid models.
4. To give examples of constructions involving the Boolean operators – **Union**, **Subtract** and **Intersect**.

Introduction

As shown in Chapter 1 the AutoCAD coordinate system includes a third coordinate direction **Z**, which, when dealing with 2D drawing in previous chapters, has not been used. 3D model drawings make use of this third **Z** coordinate.

The 3D Basics workspace

It is possible to construct 3D model drawings in the **2D Drafting & Annotation** workspaces, but in **Part 2** of this book we will be working in either the **3D Basics** or in the **3D Modeling** workspaces. To set the first of these workspaces *click* the **Workspace Settings** icon in the status bar and select **3D Introduction** from the menu which appears (Fig. 12.1). The **3D Basics** workspace appears (Fig. 12.2).

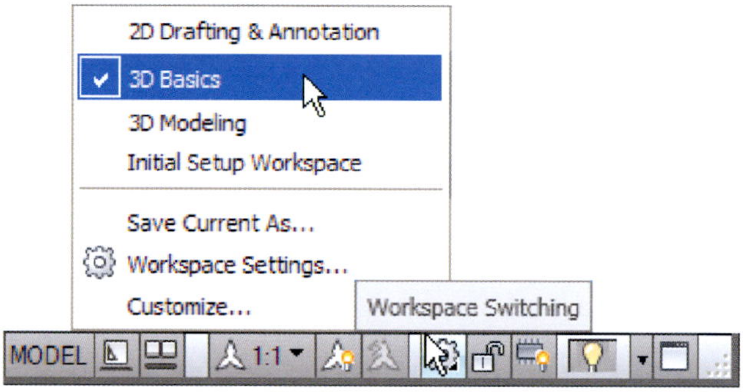

Fig. 12.1 Selecting **3D Basics** from the **Workspace Switching** menu

The workspace in Fig. 12.2 is the window in which the examples in this chapter will be constructed.

Methods of calling tools for 3D modeling

The default panels of the **3D Basics** ribbon are shown in Fig. 12.3.

When calling the tools for the construction of 3D model drawings, 3D tools can be called by:

1. A *click* on a tool icon in a **3D Basics** panel.
2. *Entering* the tool name at the command line followed by pressing the *Return* button of the mouse or the **Return** key of the keyboard.

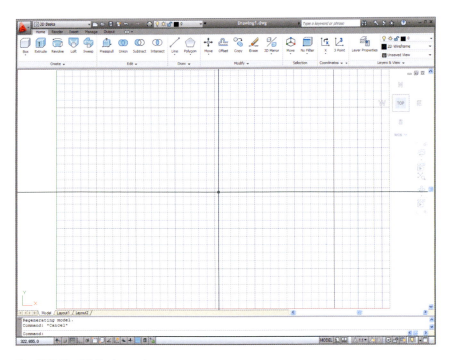

Fig. 12.2 The **3D Basics** workspace

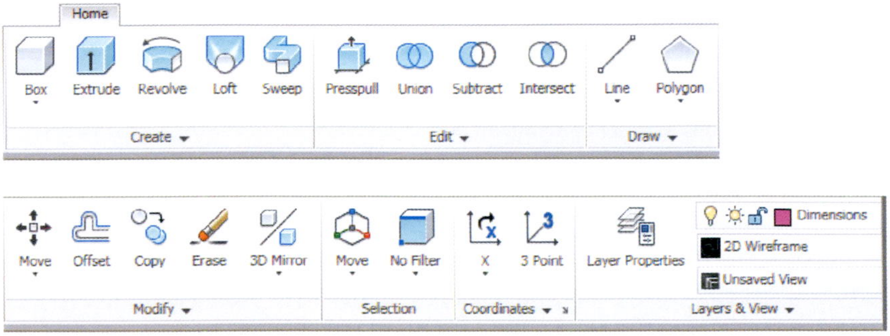

Fig. 12.3 The default **3D Basics** panels

3. Some of the 3D tools have an abbreviation which can be *entered* at the command line instead of its full name.

4. Using the **Dynamic Input** method.

Notes

1. As when constructing 2D drawings, no matter which method is used and most operators will use a variety of these four methods, the result of calling a tool results in prompt sequences appearing at the

command prompt (or if using **Dynamic Input** on screen) as in the following example:

```
Command: enter box right-click
Specify first corner or [Center]: enter 90,120
  right-click
Specify other corner or [Cube/Length]: enter
  150,200
Specify height or [2Point]: enter 50
Command:
```

Or, if the tool is called from its tool icon, or from a drop-down menu:

```
Command:_box
Specify first corner or [Center]: enter 90,120
  right-click
Specify other corner or [Cube/Length]: enter
  150,200
Specify height or [2Point]: enter 50
Command:
```

2. In the following pages, if the tool's sequences are to be repeated, they may be replaced by an abbreviated form such as:

```
Command: box
[prompts]: 90,120
[prompts]: 150,200
```

3. The examples shown in this chapter will be based on layers set as follows:

   a. *Click* the **Layer Properties** icon in the **Home/Layers & View** panel (Fig. 12.4).

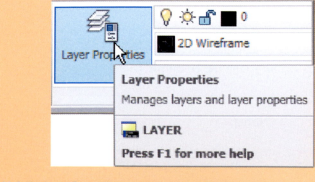

Fig. 12.4 The **Layer Properties** icon in the **Layers & View** panel

b. In the **Layer Properties Manager** which appears make settings as shown in Fig. 12.5.

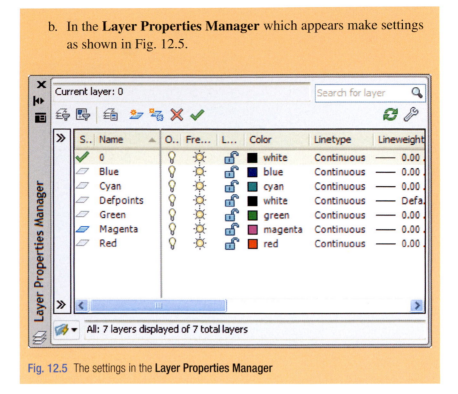

Fig. 12.5 The settings in the **Layer Properties Manager**

The Polysolid tool (Fig. 12.8)

1. Set layer **Blue** as the current layer.
2. Construct an octagon of edge length **60** using the **Polygon** tool.
3. *Click* **SW Isometric** in the **Layers & View** panel (Fig. 12.6).
4. Call the **Polysolid** tool from the **Home/Create panel** (Fig. 12.7).

The command line shows:

```
Command: _Polysolid Height=0, Width=0,
  Justification=Center
Specify start point or [Object/Height/Width/
  Justify] <Object>: enter h right-click
Specify height <0>: enter 60 right-click
Height=60, Width=0, Justification=Center
Specify start point or [Object/Height/Width/
  Justify] <Object>: enter w right-click
Specify width <0>: 5
Height=60, Width=5, Justification=Center
```

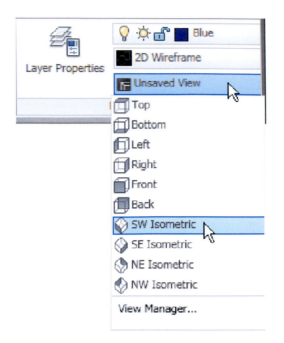

Fig. 12.6 Selecting **SW Isometric** from **3D Navigation** drop-down menu in the **Layers & View** panel

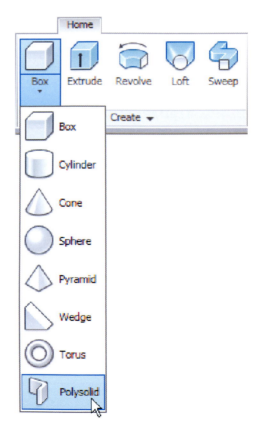

Fig. 12.7 The **Polysolid** tool icon in the **Home/Create** panel

```
Specify start point or [Object/Height/Width/
  Justify] <Object>: pick the polygon
Select object: right-click
Command:
```

And the **Polysolid** forms.

5. Select **Conceptual** from the **Layers & View** panel (Fig. 12.8).

The result is shown in Fig. 12.9.

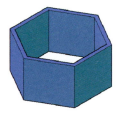

Fig. 12.9 The **Polysolid** tool example

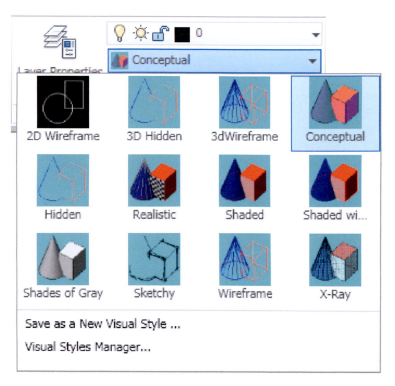

Fig. 12.8 Selecting **Conceptual** shading from **Visual Styles** in the **Layers & View** panel

2D outlines suitable for 3D models

When constructing 2D outlines suitable as a basis for constructing some forms of 3D model, select a tool from the **Home/Draw** panel, or *enter* tool names or abbreviations for the tools at the command line. If constructed using tools such as **Line**, **Circle** and **Ellipse**, before being of any use for 3D modeling, outlines must be changed into regions with the **Region** tool. Closed polylines can be used without the need to use the **Region** tool.

CHAPTER 12

Example – Outlines & Region (Fig. 12.10)

1. Construct the left-hand drawing of Fig. 12.10 using the **Line and Circle** tools.

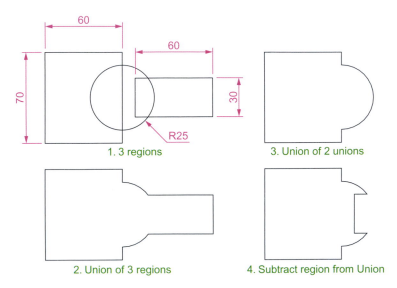

Fig. 12.10 Example – Line and circle outlines and Region

2. *Enter* **region** or **reg** at the command line. The command line shows:

```
Command:_region
Select objects: window the left-hand rectangle
  1 found
Select objects: right-click
1 loop extracted.
1 Region created.
Command:
```

And the **Line** outline is changed to a region. Repeat for the circle and the right-hand rectangle. Three regions will be formed.

3. Drawing **2** – call the **Union** tool from the **Home/Edit** panel (Fig. 12.11). The command line shows:

```
Command: _union
Select objects: pick the left-hand region 1 found
Select objects: pick the circular region 1 found,
  2 total
Select objects: pick the right-hand region 1 found,
  3 total
Select objects: right-click
Command:
```

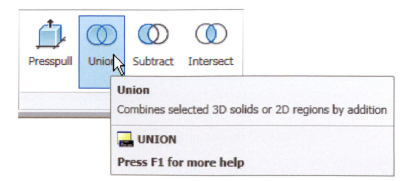

Fig. 12.11 Selecting the **Union** tool from the **Home/Edit** panel

4. Drawing **3** – with the **Union** tool form a union of the left-hand region and the circular region.
5. Drawing **4** – call the **Subtract** tool, also from the **Home/Edit** panel. The command line shows:

```
Command:_subtract Select solids and regions to
  subtract from ...
Select objects: pick the region just formed 1 found
Select objects: right-click
Select solids and regions to subtract ...
Select objects: pick the right-hand region 1 found
Select objects: right-click
Command:
```

The Extrude tool

The **Extrude** tool can be called with a *click* on its name in the **Home/Create** panel (Fig. 12.12), or by *entering* **extrude** or its abbreviation **ext** at the command line.

Fig. 12.12 The **Extrude** tool from the **Home/Create** panel

CHAPTER 12

Examples of the use of the Extrude tool

The first two examples of forming regions given in Figs 12.10 and 12.11 are used to show the results of using the **Extrude** tool.

First example – Extrude (Fig. 12.13)

From the first example of forming a region:

1. Open Fig. 12.10. Erase all but the region **2**.
2. Make layer **Green** current.
3. Call **Extrude** (Fig. 12.12). The command line shows:

```
Command: _extrude
Current wire frame density: ISOLINES=4
Closed profiles creation mode=Solid
Select objects to extrude or [MOde]: pick region
  1 found
Select objects to extrude or [MOde]: right click
Specify height of extrusion or [Direction/Path/
  Taper angle/Expression] <45>: enter 50 right-
  click
Command:
```

4. Place in the **Layers & View/3D Navigation/SW/Isometric** view.
5. Call **Zoom** and zoom to **1**.
6. Place in **Visual Style/Realistic**.

The result is shown in Fig. 12.13.

Fig. 12.13 First example – **Extrude**

Notes

1. In the above example we made use of an isometric view possible from the **3D Navigation** drop-down menu in the **Home/Layers & Views** panel (Fig. 12.6). The **3D Navigation** drop-down menu allows a model to be shown in a variety of views.

2. Note the **Current wire frame density: ISOLINES=4** in the prompts sequence when **Extrude** is called. The setting of **4** is suitable when extruding plines or regions consisting of straight lines, but when arcs are being extruded it may be better to set **ISOLINES** to a higher figure as follows:

```
Command: enter isolines right-click
Enter new value for ISOLINES <4>: enter 16
  right-click
Command:
```

3. Note the prompt **[MOde]** in the line

```
Select objects to extrude or [MOde]:
```

If **mo** is *entered* as a response to this prompt line, the following prompts appear:

```
Closed profiles creation mode[SOlid/SUrface]
  <Solid>: _SO
```

which allows the extrusion to be in solid or surface format.

Second example – Extrude (Fig. 12.14)

1. Open Fig. 12.10 and erase all but the region **3**.
2. Make the layer **Blue** current.
3. Set **ISOLINES** to **16**.
4. Call the **Extrude** tool. The command line shows:

```
Command: _extrude
Current wire frame density: ISOLINES=4, Closed
  profiles creation mode=Solid
Select objects to extrude or [MOde]: _MO Closed
  profiles creation mode
[SOlid/SUrface] <Solid>: _SO
Select objects to extrude or [MOde]: pick the
  region 3 1 found
Select objects to extrude or [MOde]:
```

CHAPTER 12

```
Specify height of extrusion or [Direction/Path/
  Taper angle/Expression]: enter t right-click
Specify angle of taper for extrusion or
  [Expression] <0>: enter 10 right-click
Specify height of extrusion or [Direction/Path/
  Taper angle/Expression]: enter 100 right-click
Command:
```

Fig. 12.14 Second example – **Extrude**

3. In the **Layers & View/3D Navigation** menu select **NE Isometric**.
4. **Zoom** to **1**.
5. Place in **Visual Styles/Hidden**.

The result is shown in Fig. 12.14.

Third example – Extrude (Fig. 12.16)

1. Make layer **Magnolia** current.
2. Construct an **80 × 50** rectangle, filleted to a radius of **15**. Then in the **3D Navigation/Front** view and using the **3D Polyline** tool from the **Home/Draw** panel (Fig. 12.15), construct **3** 3D polylines each of length **45** and at **45 degree** to each other at the centre of the outline as shown in Fig. 12.16.
3. Place the screen in the **3D Navigation/SW Isometric** view.
4. Set **ISOLINES** to **24**.
5. Call the **Extrude** tool. The command line shows:

```
Command: _extrude
Current wire frame density: ISOLINES = 24, Closed
  profiles creation mode = Solid
Select objects to extrude or [MOde]: _MO Closed
  profiles creation mode
[SOlid/SUrface] <Solid>: _SO
Select objects to extrude or [MOde]: pick the
  rectangle 1 found
Select objects to extrude or [MOde]: right-click
Specify height of extrusion or [Direction/Path/
  Taper angle/Expression]:enter t right-click
Select extrusion path or [Taper angle]: pick path
  right-click
Command:
```

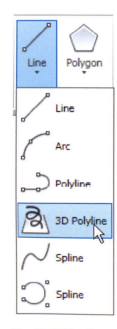

Fig. 12.15 The **3D Polyline** tool from the **Home/Draw** panel

6. Place the model in **Visual Styles/Realistic**.

The result is shown in Fig. 12.16.

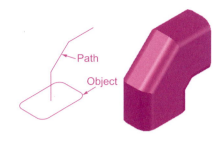

Fig. 12.16 Second example – **Extrude**

The Revolve tool

The **Revolve** tool can be called with a *click* on its tool icon in the **Home/Create** panel, by a *click* or by *entering* **revolve** at the command line, or its abbreviation **rev**.

Examples of the use of the Revolve tool

Solids of revolution can be constructed from closed plines or from regions.

First example – Revolve (Fig. 12.19)

1. Construct the closed polyline (Fig. 12.17).
2. Make layer **Red** current.
3. Set **ISOLINES** to **24**.
4. Call the **Revolve** tool from the **Home/Create** panel (Fig. 12.18).

The command line shows:

```
Command: _revolve
Current wire frame density: ISOLINES=4, Closed
  profiles creation mode=Solid
Select objects to revolve or [MOde]: _MO Closed
  profiles creation mode[SOlid/SUrface] <Solid>: _SO
```

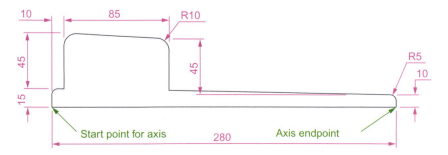

Fig. 12.17 First example – **Revolve**. The closed pline

Fig. 12.18 The **Revolve** tool from the **Home/Create** panel

```
Select objects to revolve or [MOde]: pick the
  pline 1 found
Select objects to revolve or [MOde]: right-click
Specify axis start point or define axis by [Object/
  X/Y/Z] <Object>: pick
Specify axis endpoint: pick
Specify angle of revolution or [STart angle/
  Reverse/Expression] <360>: right-click
Command:
```

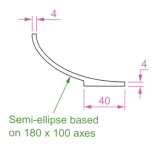

Fig. 12.19 First example – **Revolve**

5. Place in the **3D Navigation/NE Isometric** view. **Zoom** to **1**.
6. Shade with **Visual Styles/Shaded**.

The result is shown in Fig. 12.19.

Second example – Revolve (Fig. 12.21)

1. Make layer **Yellow** current.
2. Place the screen in the **3D Navigate/Front** view. **Zoom** to **1**.
3. Construct the pline outline (Fig. 12.20).

4

4

40

Semi-ellipse based
on 180 x 100 axes

Fig. 12.20 Second example – **Revolve**. The pline outline

4. Set **ISOLINES** to **24**.
5. Call the **Revolve** tool and construct a solid of revolution.
6. Place the screen in the **3D Navigate/SW Isometric. Zoom** to **1**.
7. Place in **Visual Styles/Shades of Gray** (Fig. 12.21).

Fig. 12.21 Second example – **Revolve**

Third example – Revolve (Fig. 12.22)

1. Make **Green** the current layer.
2. Place the screen in the **3D Navigate/Front** view.
3. Construct the pline (left-hand drawing of Fig. 12.22). The drawing must be either a closed pline or a region.
4. Set **Isolines** to **24**.
5. Call **Revolve** and form a solid of revolution through **180 degree**.
6. Place the model in the **3D Navigate/NE Isometric. Zoom** to **1**.
7. Place in **Visual Styles/Conceptual**.

The result is shown in Fig. 12.22 (right-hand drawing).

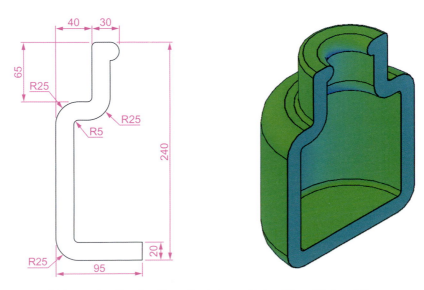

Fig. 12.22 Third example – **Revolve**. The outline to be revolved and the solid of revolution

Other tools from the Home/Create panel

First example – Box (Fig. 12.24)

1. Make **Magenta** the current layer.
2. Place the window in the **3D Navigate/Front** view.
3. Set **Isolines** to **4**.
4. *Click* the **Box** tool icon in the **Home/Create** panel (Fig. 12.23). The command line shows:

```
Command: _box
Specify first corner or [Center]: enter 90,90
  right-click
Specify other corner or [Cube/Length]: enter 110,
  -30 right-click
Specify height or [2Point]: enter 75 right-click
Command: right-click
BOX Specify first corner or [Center]: 110,90
Specify other corner or [Cube/Length]: 170,70
Specify height or [2Point]: 75
Command:
BOX Specify first corner or [Center]: 110,-10
Specify other corner or [Cube/Length]: 200,-30
Specify height or [2Point]: 75
Command:
```

Fig. 12.23 Selecting **Box** from the **Home/Create** panel

5. Place in the **ViewCube/Isometric** view. **Zoom** to **1**.
6. Call the **Union** tool from the **Home/Edit** panel. The command line shows:

```
Command:_union
Select objects: pick one of the boxes 1 found
Select objects: pick the second of box 1 found,
  2 total
Select objects: pick the third box 1 found,
  3 total
Select objects: right-click
Command:
```

And the three boxes are joined in a single union.

7. Place in **Visual Styles/Conceptual**.

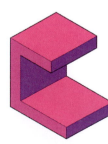

Fig. 12.24 First example – **Box**

The result is given in Fig. 12.24.

Second example – Sphere and Cylinder (Fig. 12.25)

1. Make layer **Green** current.
2. Set **ISOLINES** to **16**.
3. *Click* the **Sphere** tool icon from the **Home/Create** panel. The command line shows:

```
Command: _sphere
Specify center point or [3P/2P/Ttr]: 180,170
Specify radius or [Diameter]: 50
Command:
```

4. *Click* the **Cylinder** tool icon in the **Home/Create** panel. The command line shows:

```
Command: _cylinder
Specify center point of base or [3P/2P/Ttr/
  Elliptical]: 180,170
Specify base radius or [Diameter]: 25
Specify height or [2Point/Axis endpoint]: 110
Command:
```

5. Place the screen in the **3D Navigate/Front** view. **Zoom** to **1**.
6. With the **Move** tool (from the **Home/Modify** panel), move the cylinder vertically down so that the bottom of the cylinder is at the bottom of the sphere.
7. *Click* the **Subtract** tool icon in the **Home/Edit** panel. The command line shows:

```
Command: _subtract Select solids and regions to
  subtract from...
Select objects: pick the sphere 1 found
Select objects: right-click
Select solids and regions to subtract
Select objects: pick the cylinder 1 found
Select objects: right-click
Command:
```

8. Place the screen in **3D Navigate/SW Isometric. Zoom** to **1**.
9. Place in **Visual Styles/Realistic**.

Fig. 12.25 Second example – **Sphere and Cylinder**

The result is shown in Fig. 12.25.

Third example – Cylinder, Cone and Sphere (Fig. 12.26)

1. Make **Blue** the current layer.
2. Set **Isolines** to **24**.

3. Place in the **3D Navigate/Front** view.
4. Call the **Cylinder** tool and with a centre **170,150** construct a cylinder of radius **60** and height **15**.
5. *Click* the **Cone** tool in the **Home/Create** panel. The command line shows:

```
Command: _cone
Specify center point of base or [3P/2P/Ttr/
    Elliptical]: 170,150
Specify base radius or [Diameter]: 40
Specify height or [2Point/Axis endpoint/Top
    radius]: 150
Command:
```

6. Call the **Sphere** tool and construct a sphere of centre **170,150** and radius **45**.
7. Place the screen in the **3D Navigate/Front** view and with the **Move** tool, move the cone and sphere so that the cone is resting on the cylinder and the centre of the sphere is at the apex of the cone.
8. Place in the **3D Navigate/SW Isometric** view, **Zoom** to **1** and with **Union** form a single 3D model from the three objects.
9. Place in **Visual Styles/Conceptual**.

The result is shown in Fig. 12.26.

Fig. 12.26 Third example – **Cylinder, Cone and Sphere**

Fourth example – Box and Wedge (Fig. 12.27)

1. Make layer **Blue** current.
2. Place in the **3D Navigate/Top** view.
3. *Click* the **Box** tool icon in the **Home/Create** panel and construct two boxes, the first from corners **70,210** and **290,120** of height **10**, the second of corners **120,200,10** and **240,120,10** and of height **80**.
4. Place the screen in the **3D Navigate/Front** view and **Zoom** to **1**.
5. *Click* the **Wedge** tool icon in the **Home/Create** panel. The command line shows:

```
Command: _wedge
Specify first corner or [Center]: 120,170,10
Specify other corner or [Cube/Length]:
    80,160,10
Specify height or [2Point]: 70
Command: right-click
WEDGE
Specify first corner of wedge or [Center]:
    240,170,10
```

```
Specify corner or [Cube/Length]: 280,160,10
Specify height or [2Point]: 70
Command:
```

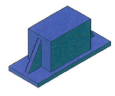

Fig. 12.27 Fourth example – **Box** and **Wedge**

6. Place the screen in **3D Navigate/SW Isometric** and **Zoom** to **1**.
7. Call the **Union** tool from the **Home/Edit** panel and in response to the prompts in the tool's sequences *pick* each of the 4 objects in turn to form a union of the 4 objects.
8. Place in **Visual Styles/Conceptual**.

The result is shown in Fig. 12.27.

Fifth example – Cylinder and Torus (Fig. 12.28)

1. Make layer **Red** current.
2. Set **Isolines** to **24**.
3. Using the **Cylinder** tool from the **Home/Create** panel, construct a cylinder of centre **180,160**, of radius **40** and height **120**.
4. *Click* the **Torus** tool icon in the **Home/Create** panel. The command line shows:

```
Command: _torus
Specify center point or [3P/2P/Ttr]:
  180,160,10
Specify radius or [Diameter]: 40
Specify tube radius or [2Point/Diameter]: 10
Command: right-click
TORUS
Specify center point or [3P/2P/Ttr]:
  180,160,110
Specify radius or [Diameter] <40>: right-click
Specify tube radius or [2Point/Diameter] <10>:
  right-click
Command:
```

5. Call the **Cylinder** tool again and construct another cylinder of centre **180, 160**, of radius **35** and height **120**.
6. Place in the **3D Navigate/SW Isometric** view and **Zoom** to **1**.
7. *Click* the **Union** tool icon in the **Home/Edit** panel and form a union of the larger cylinder and the two torii.
8. *Click* the **Subtract** tool icon in the **Home/Edit** panel and subtract the smaller cylinder from the union.
9. Place in **Visual Styles/X-Ray**.

Fig. 12.28 Fifth example – **Cylinder** and **Torus**

The result is shown in Fig. 12.28.

CHAPTER 12

The Chamfer and Fillet tools

Example – Chamfer and Fillet (Fig. 12.33)

1. Set layer **Green** as the current layer.
2. Set **Isolines** to **16**.
3. Working to the sizes given in Fig. 12.29 and using the **Box** and **Cylinder** tools, construct the 3D model (Fig. 12.30).
4. Place in the **3D Navigate/SW Isometric** view. **Union** the two boxes and with the **Subtract** tool, subtract the cylinders from the union.

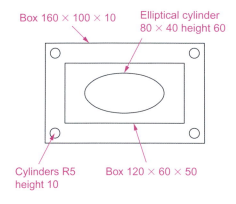

Box 160 × 100 × 10
Elliptical cylinder 80 × 40 height 60
Cylinders R5 height 10
Box 120 × 60 × 50

Fig. 12.29 Example – **Chamfer** and **Fillet** – sizes for the model

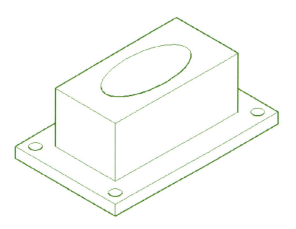

Fig. 12.30 Example – **Chamfer** and **Fillet** – isometric view – the model before using the tools

Notes

To construct the elliptical cylinder, call the **Cylinder** tool from the **Home/Modeling** panel. The command line shows:

```
Command: _cylinder
Specify center point of base or [3P/2P/Ttr/
  Elliptical]: enter e right-click
Specify endpoint of first axis or [Center]:
  130,160
Specify other endpoint of first axis: 210,160
Specify endpoint of second axis: 170,180
Specify height or [2Point/Axis endpoint]: 50
Command:
```

5. *Click* the **Fillet** tool icon in the **Home/Modify** panel (Fig. 12.31). The command line shows:

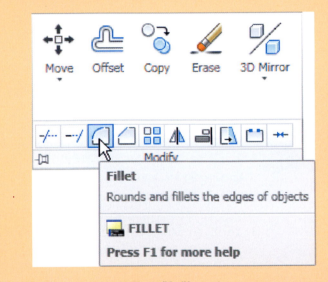

Fig. 12.31 The **Fillet** tool icon in the **Home/Modify** panel

```
Command:_fillet
Current settings: Mode=TRIM. Radius=0
Specify first object or [Undo/Polyline/Radius/
  Trim/Multiple]: enter r (Radius) right-click
Specify fillet radius <0>: 10
Select first object: pick one corner
Select an edge or [Chain/Radius]: pick a second
  corner
```

```
Select an edge or [Chain/Radius]: pick a third
  corner
Select an edge or [Chain/Radius]: pick the
  fourth corner
Select an edge or [Chain/Radius]: right-click
4 edge(s) selected for fillet.
Command:
```

6. *Click* the **Chamfer** tool in the **Home/Modify** panel (Fig. 12.32). The command line shows:

Fig. 12.32 The **Chamfer** tool icon in the **Home/Modify** panel

```
Command: _chamfer
(TRIM mode) Current chamfer Dist1 = 0, Dist2 = 0
Select first line or [Undo/Polyline/Distance/
  Angle/Trim/mEthod/Multiple]: enter d right-click
Specify first chamfer distance <0>: 10
Specify second chamfer distance <10>:
Select first line or [Undo/Polyline/Distance/
  Angle/Trim/mEthod/Multiple]: pick one corner
  One side of the box highlights
Base surface selection...
Enter surface selection option [Next/OK
  (current)] <OK>: right-click
Specify base surface chamfer distance <10>:
  right-click
Specify other surface chamfer distance <10>:
  right-click
Select an edge or [Loop]: pick the edge
Select an edge or [Loop]: pick the second edge
Select an edge [or Loop]: right-click
Command:
```

And the edges are chamfered. Repeat to chamfer the other three edges.

7. Place in **Visual Styles/Shaded with Edges**.

Fig. 12.33 shows the completed 3D model.

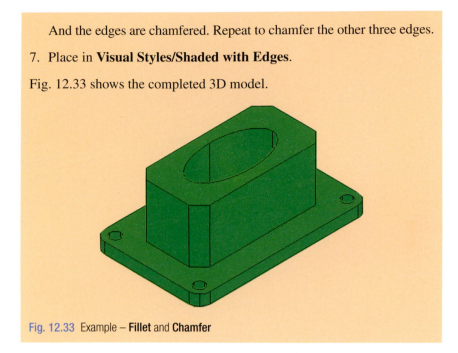

Fig. 12.33 Example – **Fillet** and **Chamfer**

Note on the tools Union, Subtract and Intersect

The tools **Union**, **Subtract** and **Intersect** found in the **Home/Edit** panel are known as the **Boolean** operators after the mathematician Boolean. They can be used to form unions, subtractions or intersection between extrusions solids of revolution, or any of the 3D Objects.

Constructing 3D surfaces using the Extrude tool

In this example of the construction of a 3D surface model the use of the **Dynamic Input** (DYN) method of construction will be shown.

1. Place the AutoCAD drawing area in the **3D Navigation/SW Isometric** view.
2. *Click* the **Dynamic Input** button in the status bar to make dynamic input active.

Example – Dynamic Input (Fig. 12.36)

1. Using the **Line** tool from the **Home/Draw** panel construct the outline (Fig. 12.34).
2. Call the **Extrude** tool and window the line outline.
3. Extrude to a height of **100**.

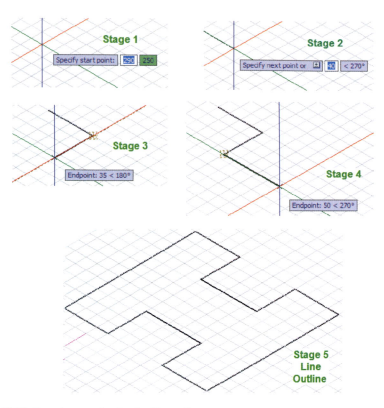

Fig. 12.34 Example – constructing the **Line** outline

The stages of producing the extrusion are shown in Figs 12.34 and 12.35. The resulting 3D model is a surface model.

> **Note**
>
> The resulting 3D model shown in Fig. 12.35 is a surface model because the extrusion was constructed from an outline consisting of lines, which are individual objects in their own right. If the outline had been a polyline, the resulting 3D model would have been a solid model. The setting of **MOde** makes no difference.

The Sweep tool

To call the tool *click* on its tool icon in the **Home/Create** panel (Fig. 12.36).

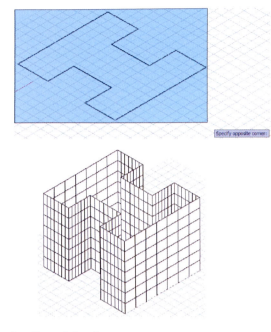

Fig. 12.35 Example – **Dynamic Input**

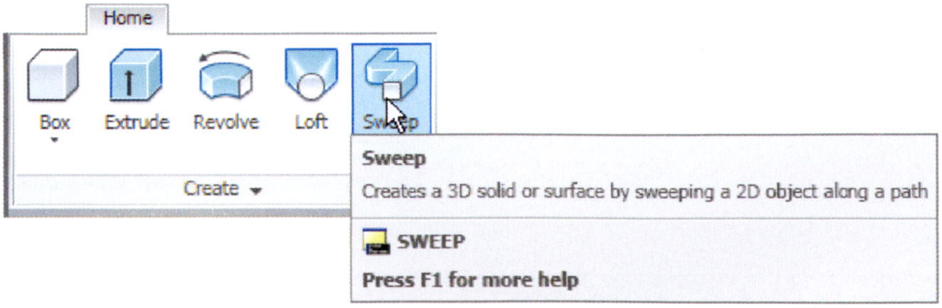

Fig. 12.36 Selecting the **Sweep** tool from the **Home/Create** panel

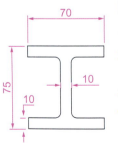

Fig. 12.37 Example **Sweep** – the outline to be swept

Example – Sweep (Fig. 12.38)

1. Construct the pline outline (Fig. 12.37) in the **3D Navigation/Top** view.
2. Change to the **3D Navigation/Front** view, **Zoom** to **1** and construct a pline as shown in Fig. 12.38 as a path central to the outline.
3. Make the layer **Magenta** current.
4. Place the window in the **3D Navigation/SW Isometric** view and *click* the **Sweep** tool icon. The command line shows:

```
Command: _sweep
Current wire frame density: ISOLINES=4, Closed
  profiles creation mode=Solid
```

```
Select objects to sweep or [MOde]: _MO Closed
  profiles creation mode
[SOlid/SUrface] <Solid>: _SO
Select objects to sweep or [MOde]: pick the pline
  1 found
Select objects to sweep or [MOde]: right-click
Select sweep path or [Alignment/Base point/Scale/
  Twist]: pick the pline path
Command:
```

5. Place in **Visual Styles/Shaded**.

The result is shown in Fig. 12.38.

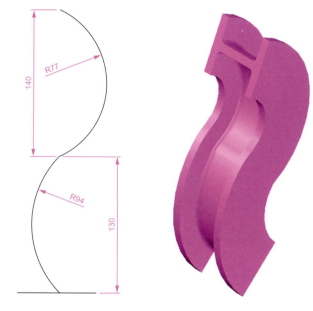

Fig. 12.38 Example – **Sweep**

The Loft tool

To call the tool *click* on its icon in the **Home/Create** panel.

Example – Loft (Fig. 12.41)

1. In the **3D Navigate/Top** view, construct the seven circles shown in Fig. 12.39 at vertical distances of **30** units apart.
2. Place the drawing area in the **3D Navigate/SW Isometric** view.
3. Call the **Loft** tool with a *click* on its tool icon in the **Home/Modeling** panel (Fig. 12.40).

CHAPTER 12

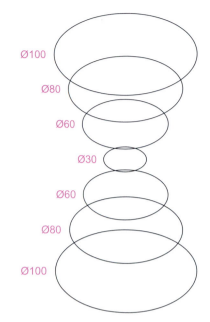

Ø100
Ø80
Ø60
Ø30
Ø60
Ø80
Ø100

Fig. 12.39 Example **Loft** – the cross sections

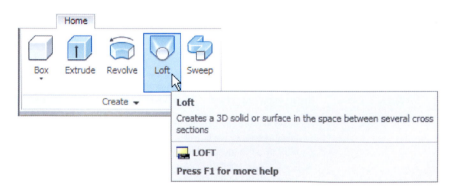

Fig. 12.40 Selecting the **Loft** tool from the **Home/Create** panel

4. Set **Cyan** as the current layer.
5. The command line shows:

```
Command:_loft
Select cross sections in lofting order or
  [POint/Join multiple curves]: pick 1 found
Select cross sections in lofting order or [POint/
  Join multiple curves]: pick 1
found, 2 total
Select cross sections in lofting order or [POint/
  Join multiple curves]: pick 1
```

```
found, 3 total
Select cross sections in lofting order or [POint/
   Join multiple curves]: pick 1 found, 4 total
Select cross sections in lofting order or [POint/
   Join multiple curves]: pick 1 found, 5 total
Select cross sections in lofting order or [POint/
   Join multiple curves]: pick 1 found, 6 total
Select cross sections in lofting order or [POint/
   Join multiple curves]: pick 1 found, 7 total
Select cross sections in lofting order or [POint/
   Join multiple curves]: enter j right-click
Select curves that are to be joined into a single
   cross section: right-click 7 cross sections
   selected
Enter an option [Guides/Path/Cross sections only/
   Settings] <Cross sections only>: right-click
Command:
```

Fig. 12.41 Example –
Loft

6. Place in **Visual Styles/Shaded with Edges**.

The result is shown in Fig. 12.41.

REVISION NOTES

1. In the AutoCAD 3D coordinate system, positive Z is towards the operator away from the monitor screen.
2. A 3D face is a mesh behind which other details can be hidden.
3. The Extrude tool can be used for extruding closed plines or regions to stated heights, to stated slopes or along paths.
4. The Revolve tool can be used for constructing solids of revolution through any angle up to 360 degree.
5. 3D models can be constructed from Box, Sphere, Cylinder, Cone, Torus and Wedge. Extrusions and/or solids of revolutions may form part of models constructed using these 3D tools.
6. The tools Union, Subtract and Intersect are known as the Boolean operators.
7. When polylines form an outline which is not closed are acted upon by the Extrude tool the resulting models will be 3D Surface models irrespective of the MOde setting.

Exercises

Methods of constructing answers to the following exercises can be found in the free website:

http://books.elsevier.com/companions/978-0-08-096575-8

The exercises which follow require the use of tools from the **Home/Create** panel in association with tools from other panels.

1. Fig. 12.42 shows the pline outline from which the polysolid outline (Fig. 12.43) has been constructed to a height of **100** and **Width** of **3**. When the polysolid has been constructed, construct extrusions which can then be subtracted from the polysolid. Sizes of the extrusions are left to your judgement.

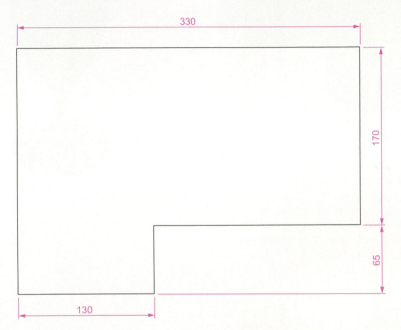

Fig. 12.42 Exercise 1 – outline for polyline

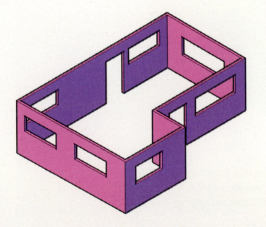

Fig. 12.43 Exercise 1

2. Fig. 12.44 shows a 3D model constructed from four polysolids which have been formed into a union using the **Union** tool from the **Home/Modify** panel. The original polysolid was formed from a hexagon of edge length **30**. The original polysolid was of height **40** and **Width 5**. Construct the union.

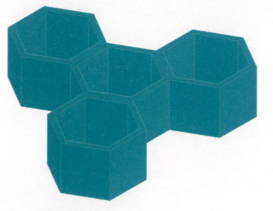

Fig. 12.44 Exercise 2

3. Fig. 12.45 shows the 3D model from Exercise 2 acted upon by the **Presspull** tool from the **Home/Create** panel.
    With the 3D model from Exercise 2 on screen and using the **Presspull** tool, construct the 3D model shown in Fig. 12.45. The distance of the pull can be estimated.

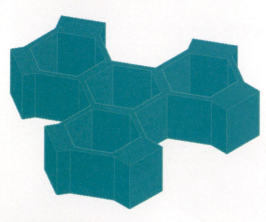

Fig. 12.45 Exercise 3

4. Construct the 3D model of a wine glass as shown in Fig. 12.46, working to the dimensions given in the outline drawing Fig. 12.47.

You will need to construct the outline and change it into a region before being able to change the outline into a solid of revolution using the **Revolve** tool from the **Home/Create** panel. This is because the semi-elliptical part of the outline has been constructed using the **Ellipse** tool, resulting in part of the outline being a spline, which cannot be acted upon by **Polyline Edit** to form a closed pline.

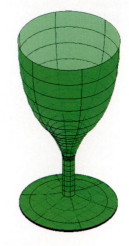

Fig. 12.46 Exercise 4

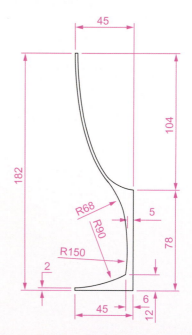

Fig. 12.47 Exercise 4 – outline drawing

5. Fig. 12.48 shows the outline from which a solid of revolution can be constructed. Using the **Revolve** tool from the **Home/Create** panel to construct the solid of revolution.

6. Construct a 3D solid model of a bracket working to the information given in Fig. 12.49.

7. Working to the dimensions given in Fig. 12.50 construct an extrusion of the plate to a height of **5** units.

8. Working to the details given in the orthographic projection (Fig. 12.51), construct a 3D model of the assembly. After

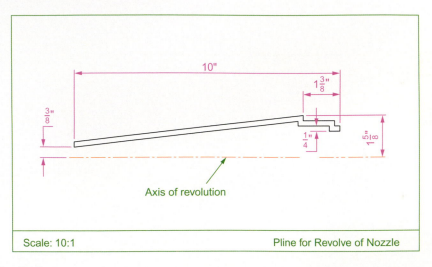

Fig. 12.48 Exercise 5

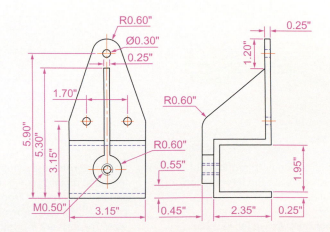

Fig. 12.49 Exercise 6

CHAPTER 12

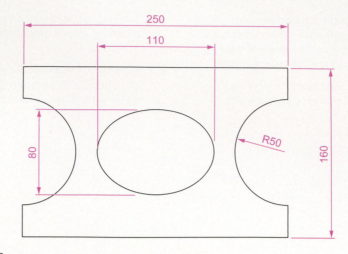

Fig. 12.50 Exercise 7

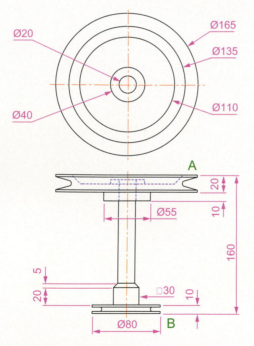

Detail at A (Scale 2:1)

Detail at B (Scale 2:1)

Fig. 12.51 Exercise 8

constructing the pline outline(s) required for the solid(s) of revolution, use the **Revolve** tool to form the 3D solid.

9. Working to the polylines shown in Fig. 12.52 construct the **Sweep** shown in Fig. 12.53.

10. Construct the cross sections as shown in the left-hand drawing of Fig. 12.54 working to suitable dimensions. From the cross sections construct the lofts shown in the right-hand view. The lofts are topped with a sphere constructed using the **Sphere** tool.

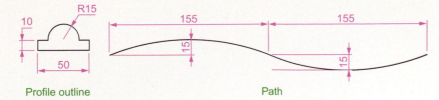

Profile outline   Path

Fig. 12.52 Exercise 9 – profile and path dimensions

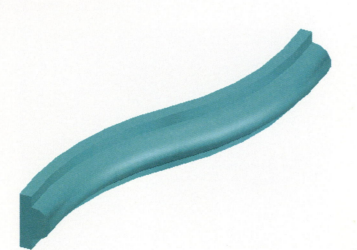

Fig. 12.53 Exercise 9

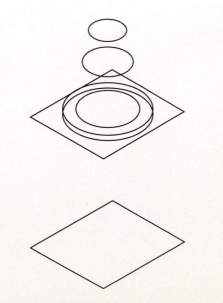

Fig. 12.54 The cross sections for Exercise 10

3D models in viewports

AIM OF THIS CHAPTER

The aim of this chapter is to give examples of 3D solid models constructed in multiple viewport settings.

The 3D Modeling workspace

In Chapter 12 all 3D model actions were constructed in the **3D Basics** workspace. As shown in that chapter, a large number of different types of 3D models can be constructed in that workspace. In the following chapters 3D models will be constructed in the **3D Modeling** workspace, brought to screen with a *click* on **3D Modeling** icon the **Workspace Settings** menu (Fig. 13.1). The AutoCAD window assumes the selected workspace settings (Fig. 13.2).

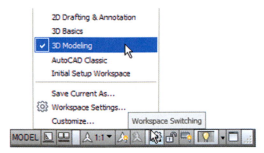

Fig. 13.1 Opening the **3D Modeling** workspace

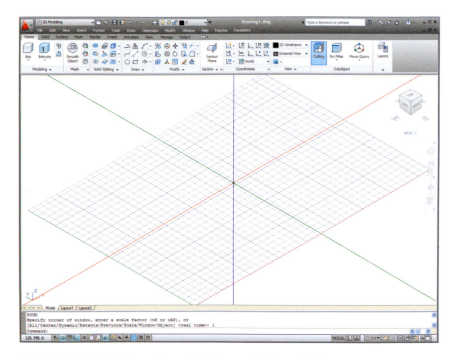

Fig. 13.2 The **3D Modeling** workspace in **SW Isometric** view and **Grid** on

If the **3D Modeling** workspace is compared with the **3D Basics** workspace (Fig. 12.2, page 225) it will be seen that there are several new tabs which when *clicked* bring changes in the ribbon with different sets of panels. In Fig. 13.2 the menu bar is included. This need not be included if the operator does not need the drop-down menus available from the menu bar.

Setting up viewport systems

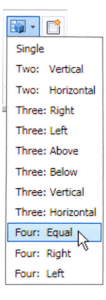

One of the better methods of constructing 3D models is in different multiple viewports. This allows what is being constructed to be seen from a variety of viewing positions. To set up multiple viewports.

In the **3D Modeling** workspace *click* **New** in the **View/Viewports** panel. From the popup list which appears (Fig. 13.3) select **Four: Equal**. The **Four: Equal** viewports layout appears (Fig. 13.4).

Fig. 13.3 Selecting **Four: Equal** from the **View/Viewports** popup list

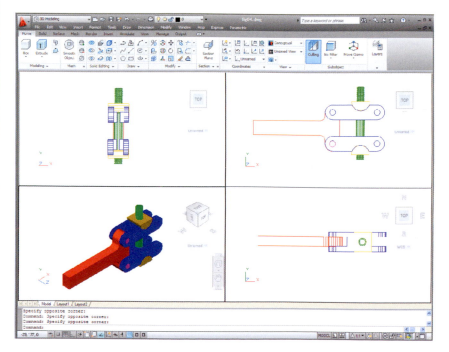

Fig. 13.4 The **Four: Equal** viewports layout

In Fig. 13.4 a simple 3D model has been constructed in the **Four: Equal** viewport layout. It will be seen that each viewport has a different view of the 3D model. Top right is an **isometric** view. Bottom right is a view from the **right** of the model. Bottom left is a view from the **left** of the model.

CHAPTER 13

Top left is a view from the **top** of the model. Note that the **front view** viewport is surrounded by a thicker line than the other three, which means it is the **current** viewport. Any one of the four viewports can be made current with a *left-click* within its boundary. Note also that three of the views are in **third angle** projection.

When a viewport system has been opened it will usually be necessary to make each viewport current in turn and **Zoom** and **Pan** to ensure that views fit well within their boundaries.

If a **first angle** layout is needed it will be necessary to open the **Viewports** dialog (Fig. 13.5) with a *click* on the **New** icon in the **View/Viewports** panel (Fig. 13.6). First select **Four: Equal** from the **Standard viewports** list; select **3D** from the **Setup** popup menu; *click* in the **top right** viewport and select **Left** in the **Change View** popup list; *enter* **first angle** in the **New name** field. Change the other viewports as shown. Save the settings with a *click* on the **Named Viewports** tab and *enter* the required name for the setup in the sub-dialog which appears.

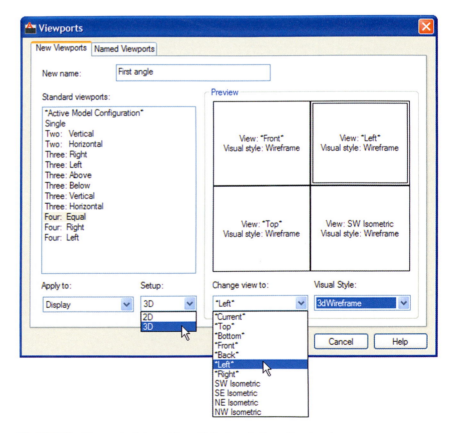

Fig. 13.5 The **Viewports** dialog set for a **3D first angle Four: Equal** setting

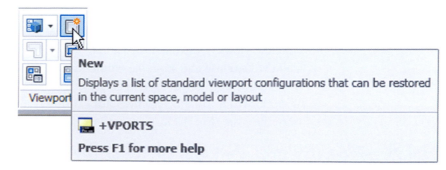

Fig. 13.6 Selecting **New from the** View/Viewports panel

First example – **Four: Equal viewports** (Fig. 13.9)

Fig. 13.7 shows a two-view orthographic projection of a support. To construct a **Scale 1:1** third angle 3D model of the support in a **Four Equal** viewport setting on a layer colour **Blue**:

1. Open a **Four Equal** viewport setting from the **New** popup list in the **View/Viewports** panel (Fig. 13.3).
2. *Click* in each viewport in turn, making the selected viewport active, and **Zoom** to **1**.

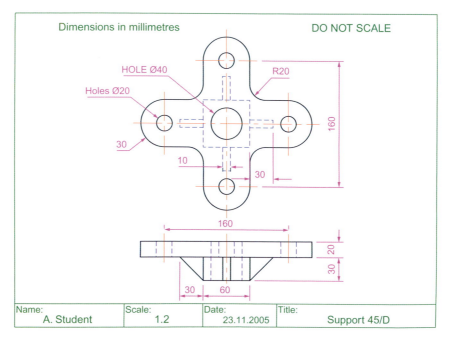

Fig. 13.7 First example – orthographic projection of the support

3. Using the **Polyline** tool, construct the outline of the plan view of the plate of the support, including the holes in the **Top** viewport (Fig. 13.5). Note the views in the other viewports.

4. Call the **Extrude** tool from the **Home/Modeling** panel and extrude the plan outline and the circles to a height of **20**.

5. With **Subtract** from the **Home/Solid Editing** panel, subtract the holes from the plate (Fig. 13.8).

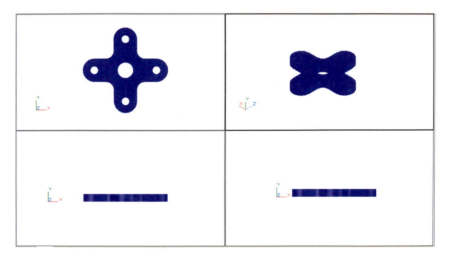

Fig. 13.8 First example – the four viewports after **Extrude** and **Subtract**

6. Call the **Box** tool and in the centre of the plate construct a box of Width=**60**, Length=**60** and Height=**30**.

7. Call the **Cylinder** tool and in the centre of the box construct a cylinder of Radius=**20** and of Height=**30**.

8. Call **Subtract** and subtract the cylinder from the box.

9. *Click* in the **Right** viewport, with the **Move** tool, move the box and its hole into the correct position with regard to the plate.

10. With **Union**, form a union of the plate and box.

11. *Click* in the **Front** viewport and construct a triangle of one of the webs attached between the plate and the box. With **Extrude**, extrude the triangle to a height of **10**. With the **Mirror** tool, mirror the web to the other side of the box.

12. *Click* in the **Right** viewport and with the **Move** tool, move the two webs into their correct position between the box and plate. Then, with **Union**, form a union between the webs and the 3D model.

13. In the **Right** viewport, construct the other two webs and in the **Front** viewport, move, mirror and union the webs as in steps **11** and **12**.

Fig. 13.9 shows the resulting four-viewport scene.

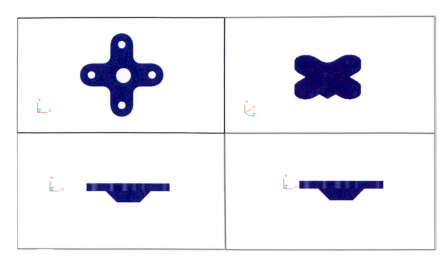

Fig. 13.9 First example – **Four: Equal viewports**

Second example – Four: Left viewports (Fig. 13.11)

1. Open a **Four: Left** viewport layout from the **Views/Viewports** popup list (Fig. 13.3).
2. Make a new layer of colour **Magenta** and make that layer current.
3. In the **Top** viewport construct an outline of the web of the Support Bracket shown in Fig. 13.10. With the **Extrude** tool, extrude the parts of the web to a height of **20**.
4. With the **Subtract** tool, subtract the holes from the web.

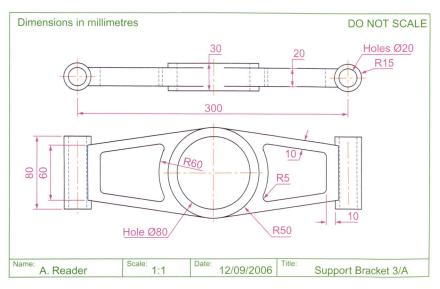

Fig. 13.10 Working drawing for the second example

5. In the **Top** viewport, construct two cylinders central to the extrusion, one of radius 50 and height 30, the second of radius 40 and height 30. With the **Subtract** tool, subtract the smaller cylinder from the larger.
6. *Click* in the **Front** viewport and move the cylinders vertically by **5** units. With **Union** form a union between the cylinders and the web.
7. Still in the **Front** viewport and at one end of the union, construct two cylinders, the first of radius **10** and height **80**, the second of radius **15** and height **80**. Subtract the smaller from the larger.
8. With the **Mirror** tool, mirror the cylinders to the other end of the union.
9. Make the **Top** viewport current and with the **Move** tool, move the cylinders to their correct position at the ends of the union. Form a union between all parts on screen.
10. Make the **Isometric** viewport current. From the **View/Visual Styles** panel select **Conceptual**.

Fig. 13.11 shows the result.

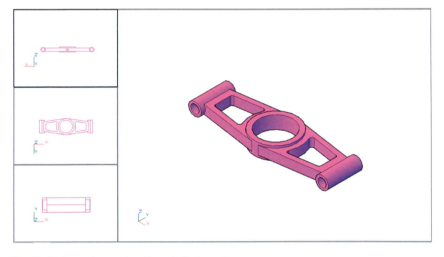

Fig. 13.11 Second example – **Four: Left** viewports

Third example – Three: Right viewports (Fig. 13.13)

1. Open the **Three: Right** viewport layout from the **View/Viewports** popup list (Fig. 13.3).
2. Make a new layer of colour **Green** and make that layer current.
3. In the **Front** viewport (top left-hand), construct a pline outline to the dimensions in Fig. 13.12.
4. Call the **Revolve** tool from the **Home/Modeling** panel and revolve the outline through 360 degree.
5. From the **View/Visual Styles** panel select **Conceptual**.

The result is shown in Fig. 13.13.

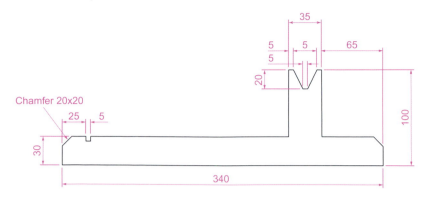

Fig. 13.12 Third example – outline for solid of revolution

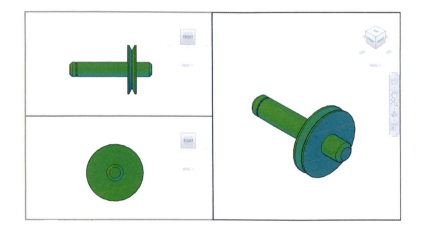

Fig. 13.13 Third example – **Three: Right viewports**

Notes

1. When working in viewport layouts, make good use of the **Zoom** tool, because the viewports are smaller than a single viewport in AutoCAD 2011.

2. As in all other forms of constructing drawings in AutoCAD 2011 frequent toggling of **SNAP**, **ORTHO** and **GRID** will allow speedier and more accurate working.

REVISION NOTES

1. Outlines suitable for use when constructing 3D models can be constructed using the 2D tools such as Line, Arc, Circle and polyline. Such outlines must either be changed to closed polylines or to regions before being incorporated in 3D models.
2. The use of multiple viewports can be of value when constructing 3D models in that various views of the model appear enabling the operator to check the accuracy of the 3D appearance throughout the construction period.

Exercises

Methods of constructing answers to the following exercises can be found in the free website: http://books.elsevier.com/companions/978-0-08-096575-8

1. Using the **Cylinder**, **Box**, **Sphere**, **Wedge** and **Fillet** tools, together with the **Union** and **Subtract** tools and working to any sizes thought suitable, construct the 'head' as shown in the **Three: Right** viewport as shown in Fig. 13.12 (Fig. 13.14).

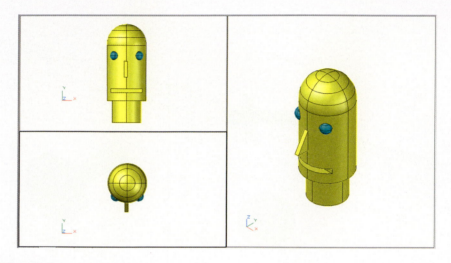

Fig. 13.14 Exercise 1

2. Using the tools **Sphere**, **Box**, Union and **Subtract** and working to the dimensions given in Fig. 13.15, construct the 3D solid model as shown in the isometric drawing Fig. 13.16.

Fig. 13.16 Exercise 2

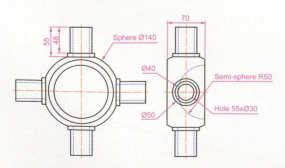

Fig. 13.15 Exercise 2 – working drawing

3. Each link of the chain shown in Fig. 13.17 has been constructed using the tool **Extrude** and extruding a small circle along an elliptical path. Copies of the link were then made, half

of which were rotated in a **Right** view and then moved into their position relative to the other links. Working to suitable sizes construct a link and from the link construct the chain as shown.

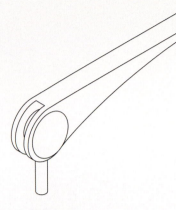

Fig. 13.19 Exercise 4

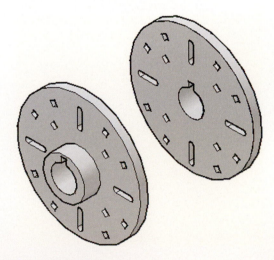

Fig. 13.17 Exercise 3

4. A two-view orthographic projection of a rotatable lever from a machine is given in Fig. 13.18 together with an isometric drawing of the 3D model constructed to the details given in the drawing Fig. 13.19.

5. Construct the 3D model drawing in a **Four: Equal** viewport setting.

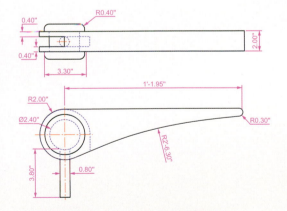

Fig. 13.18 Exercise 4 – orthographic projection

6. Working in a **Three: Left** viewport setting, construct a 3D model of the faceplate to the dimensions given in Fig. 13.20. With the **Mirror** tool, mirror the model to obtain

an opposite facing model. In the **Isometric** viewport call the **Hide** tool (Fig. 13.21).

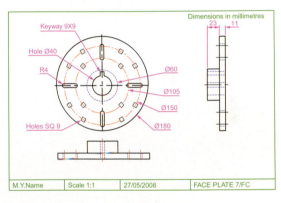

Fig. 13.20 Exercise 5 – dimensions

Fig. 13.21 Exercise 5

CHAPTER 13

The modification of 3D models

AIMS OF THIS CHAPTER

The aims of the chapter are:

1. To demonstrate how 3D models can be saved as blocks for insertion into other drawings via the **DesignCenter**.
2. To show how a library of 3D models in the form of blocks can be constructed to enable the models to be inserted into other drawings.
3. To give examples of the use of the tools from the **Home/Modify** panel:
   3D Array – Rectangular and Polar 3D arrays;
   3D Mirror;
   3D Rotate.
4. To give examples of the use of the **Helix** tool.
5. To give an example of construction involving **Dynamic Input**.
6. To show how to obtain different views of 3D models in 3D space using the **View/Views/3D Manager** and the **ViewCube**.
7. To give simple examples of surfaces using **Extrude**.

Creating 3D model libraries

In the same way as 2D drawings of parts such as electronics symbols, engineering parts, building symbols and the like can be saved in a file as blocks and then opened into another drawing by *dragging* the appropriate block drawing from the DesignCenter, so can 3D models.

First example – inserting 3D blocks (Fig. 14.4)

1. Construct 3D models of the parts for a lathe milling wheel holder to details as given in Fig. 14.1 each on a layer of different colours.

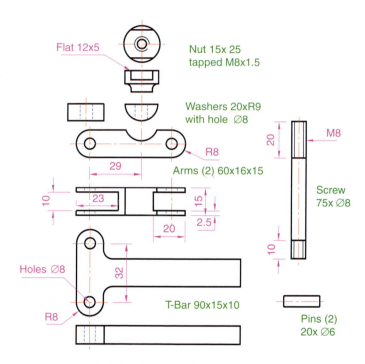

Fig. 14.1 The components of a lathe milling wheel holder

2. Save each of the 3D models of the parts to file names as given in Fig. 14.1 as blocks using **Create** from the **Insert/Block** panel. Save all seven blocks and delete the drawings on screen. Save the drawing with its blocks to a suitable file name (Fig01.dwg).
3. Set up a **Four: Equal** viewports setting.
4. Open the **DesignCenter** from the **View/Palettes** panel (Fig. 14.2) or by pressing the **Ctrl** and **2** keys of the keyboard.

Fig. 14.2 Calling the **DesignCenter** from the **View/Palettes** panel

5. In the **DesignCenter** *click* the directory **Chap14**, followed by another *click* on **Fig04.dwg** and yet another *click* on **Blocks**. The saved blocks appear as icons in the right-hand area of the **DesignCenter**.

6. *Drag* and *drop* the blocks one by one into one of the viewports on screen. Fig. 14.3 shows the **Nut** block ready to be *dragged* into the **Right** viewport. As the blocks are *dropped* on screen, they will need moving into their correct positions in suitable viewports using the **Move** tool from the **Home/Modify** panel.

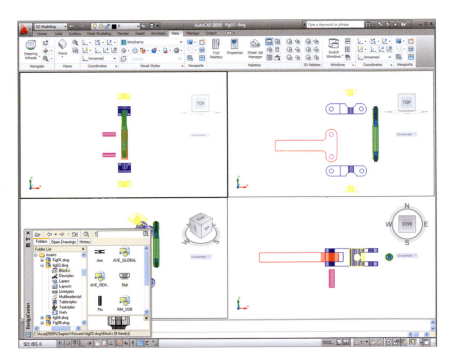

Fig. 14.3 First example – **inserting 3D blocks**

7. Using the **Move** tool, move the individual 3D models into their final places on screen and shade the **Isometric** viewport using **Conceptual** shading from the **Home/View** panel (Fig. 14.4).

CHAPTER 14

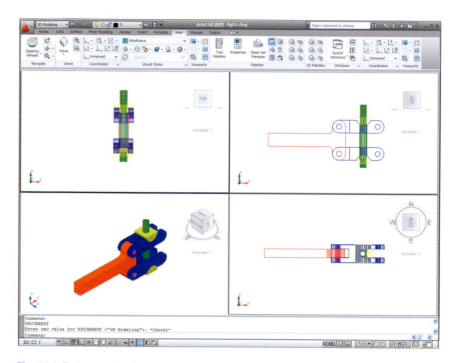

Fig. 14.4 First example – **inserting 3D blocks**

Notes

1. It does not matter in which of the four viewports any one of the blocks is *dragged* and *dropped* into. The part automatically assumes the view of the viewport.

2. If a block destined for layer **0** is *dragged* and *dropped* into the layer **Centre** (which in our **acadiso.dwt** is of colour **red** and of linetype **CENTER2**), the block will take on the colour (red) and linetype of that layer (**CENTER2**).

3. In this example, the blocks are 3D models and there is no need to use the **Explode** tool option.

4. The examples of a **Four: Equal** viewports screen shown in Figs 14.3 and 14.4 are in **first** angle. The front view is top right; the end view is top left; the plan is bottom right.

Second example – a library of fastenings (Fig. 14.6)

1. Construct 3D models of a number of engineering fastenings. In this example only five have been constructed – a 10 mm round head rivet, a 20 mm countersunk head rivet, a cheese head bolt, a countersunk head

Fig. 14.5 Second example – the five fastenings

bolt and a hexagonal head bolt together with its nut (Fig. 14.5). With the **Create** tool save each separately as a block, erase the original drawings and save the file to a suitable file name – in this example Fig05.dwg.

2. Open the DesignCenter, *click* on the **Chapter 14** directory, followed by a *click* on **Fig05.dwg**. Then *click* again on **Blocks** in the content list of **Fig05.dwg**. The five 3D models of fastenings appear as icons in the right-hand side of the DesignCenter (Fig. 14.6).

3. Such blocks of 3D models can be *dragged* and *dropped* into position in any engineering drawing where the fastenings are to be included.

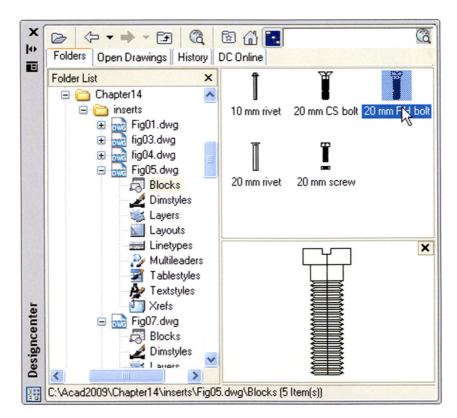

Fig. 14.6 Second example – a library of fastenings

Constructing a 3D model (Fig. 14.9)

A three-view projection of a pressure head is shown in Fig. 14.7. To construct a 3D model of the head:

1. Select **Front** from the **View/Views** panel.
2. Construct the outline to be formed into a solid of revolution (Fig. 14.8) on a layer colour magenta and with the **Revolve** tool, produce the 3D model of the outline.

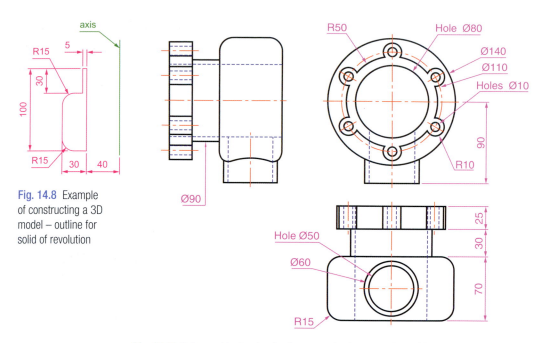

Fig. 14.8 Example of constructing a 3D model – outline for solid of revolution

Fig. 14.7 Orthographic drawing for the example of constructing a 3D model

3. Set the **View/Views/Top** view and with the **Cylinder** tool, construct cylinders as follows:
   In the centre of the solid – radius **50** and height **50**.
   With the same centre – radius **40** and height **40**. Subtract this cylinder from that of radius **50**.
   At the correct centre – radius **10** and height **25**.
   At the same centre – radius **5** and height **25**. Subtract this cylinder from that of radius **10**.

4. With the **Array** tool, form a polar **6** times array of the last two cylinders based on the centre of the 3D model.

5. Set the **View/Views/Front** view.

6. With the **Move** tool, move the array and the other two cylinders to their correct positions relative to the solid of revolution so far formed.

7. With the **Union** tool form a union of the array and other two solids.

8. Set the **View/Views/Right** view.

9. Construct a cylinder of radius **30** and height **25** and another of radius **25** and height **60** central to the lower part of the 3D solid so far formed.

10. Set the **View/Views/Top** view and with the **Move** tool move the two cylinders into their correct position.

CHAPTER 14

11. With **Union**, form a union between the radius **30** cylinder and the 3D model and with **Subtract**, subtract the radius **25** cylinder from the **3D** model.

12. *Click* **Realistic** in the **View/Visual Styles** panel list.

The result is given in Fig. 14.9.

Fig. 14.9 Example of constructing a 3D model

Notes

This 3D model could equally as well have been constructed in a three or four viewports setting. **Full Shading** has been set on from the **Render** ribbon, hence the line of shadows.

The 3D Array tool

First example – a Rectangular Array (Fig. 14.12)

1. Construct the star-shaped pline on a layer colour green (Fig. 14.10) and extrude it to a height of **20**.

2. *Click* on the **3D Array** in the **Home/Modify** panel (Fig. 14.11). The command line shows:

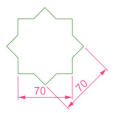

Fig. 14.10 Example – **3D Array** – the star pline

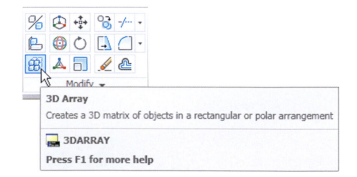

Fig. 14.11 Selecting **3D Array** from the **Home/Modify** panel

```
Command:_3darray
Select objects: pick the extrusion 1 found
Select objects: right-click
Enter the type of array [Rectangular/Polar] <R>:
  right-click
Enter the number of rows (---) <1>: enter 3 right-
  click
Enter the number of columns (III): enter 3 right-
  click
Enter the number of levels (...): enter 4 right-click
```

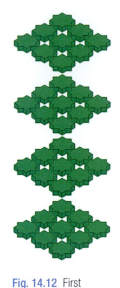

```
Specify the distance between rows (—): enter 100
  right-click
Specify the distance between columns (III): enter
  100 right-click
Specify the distance between levels (...): enter
  300 right-click
Command:
```

3. Place the screen in the **View/Views/SW Isometric** view.
4. Shade using the **View/Visual Styles/Shaded with Edges** visual style (Fig. 14.12).

Fig. 14.12 First example – a **3D Rectangular Array**

Second example – a Polar Array (Fig. 14.13)

1. Use the same star-shaped 3D model.
2. Call the **3D Array** tool again. The command line shows:

```
Command:_3darray
Select objects: pick the extrusion 1 found
Select objects: right-click
Enter the type of array [Rectangular/Polar] <R>:
  enter p (Polar) right-click
Enter number of items in the array: 12
Specify the angle to fill (+=ccw), -=cw) <360>:
  right-click
Rotate arrayed objects? [Yes/No] <Y>: right-click
Specify center point of array: 235,125
Specify second point on axis of rotation: 300,200
Command:
```

Fig. 14.13 Second example – a **3D Polar Array**

3. Place the screen in the **View/Views/SW Isometric** view.
4. Shade using the **View/Visual Styles Shaded** visual style (Fig. 14.13).

Third example – a Polar Array (Fig. 14.15)

Fig. 14.14 Third example – a **3D Polar Array** – the 3D model to be arrayed

1. Working on a layer of colour **red**, construct a solid of revolution in the form of an arrow to the dimensions as shown in Fig. 14.14.
2. *Click* **3D Array** in the **Home/Modify** panel. The command line shows:

```
Command: _3darray
Select objects: pick the arrow 1 found
Select objects: right-click
Enter the type of array [Rectangular/Polar]<R>:
  enter p right-click
Enter the number of items in the array: enter 12
  right-click
Specify the angle to fill (+=ccw, -=cw) <360>:
  right-click
Rotate arrayed objects? [Yes/No] <Y>:
  right-click
Specify center point of array: enter 40,170,20
  right-click
Specify second point on axis of rotation: enter
  60,200,100 right-click
Command:
```

3. Place the array in the **3D Navigate/SW Isometric** view and shade to **View/Visual Styles/Shades of Gray**. The result is shown in Fig. 14.15.

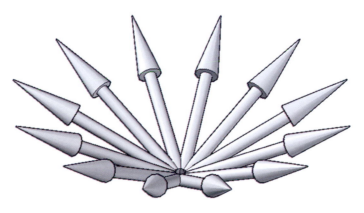

Fig. 14.15 Third example – a **3D Polar Array**

CHAPTER 14

The 3D Mirror tool

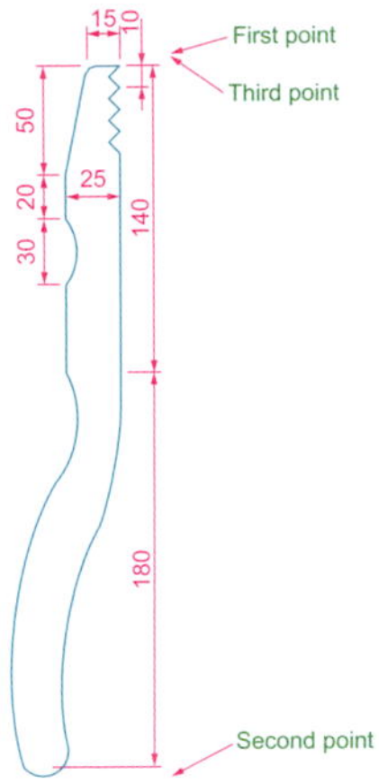

First example – 3D Mirror (Fig. 14.17)

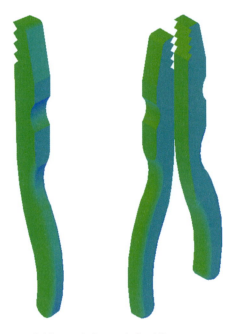

1. Working on a layer colour green, construct the outline Fig. 14.16.
2. Extrude the outline to a height of **20**.
3. Extrude the region to a height of **5** and render. A **Conceptual** style shading is shown in Fig. 14.17 (left-hand drawing).

Fig. 14.16 First example – **3D Mirror** – outline of object to be mirrored

Fig. 14.17 First example – **3D Mirror** – before and after **Mirror**

4. *Click* on **3D Mirror** in the **3D Operation** sub-menu of the **Modify** drop-down menu. The command line shows:

```
Command:_3dmirror
Select objects: pick the extrusion 1 found
Select objects: right-click
Specify first point of mirror plane (3 points): pick
Specify second point on mirror plane: pick
Specify third point on mirror plane or [Object/
  Last/Zaxis/View/XY/YZ/ZX/3points]: enter .xy
  right-click of (need Z): enter 1 right-click
Delete source objects? [Yes/No]: <N>: right-click
Command:
```

The result is shown in the right-hand illustration of Fig. 14.17.

Second example – 3D Mirror (Fig. 14.19)

1. Construct a solid of revolution in the shape of a bowl in the **3D Navigate/ Front** view working on a layer of colour **magenta** (Fig. 14.18).

Fig. 14.18 Second example – **3D Mirror** – the 3D model

2. *Click* **3D Mirror** in the **Home/Modify** panel. The command line shows:

```
Command:_3dmirror
Select objects: pick the bowl 1 found
Select objects: right-click
Specify first point on mirror plane (3 points):
  pick
Specify second point on mirror plane: pick
Specify third point on mirror plane: enter .xy
  right-click (need Z): enter 1 right-click
Delete source objects:? [Yes/No]: <N>: right-click
Command:
```

The result is shown in Fig. 14.19.

3. Place in the **3D Navigate/SW Isometric** view.
4. Shade using the **View/Visual Styles Conceptual** visual style (Fig. 14.19).

Fig. 14.19 Second example – **3D Mirror** – the result in a front view

CHAPTER 14

The 3D Rotate tool

Example – 3D Rotate (Fig. 14.20)

1. Use the same 3D model of a bowl as for the last example. *Pick* **3D Rotate** tool from the **Home/Modify** panel. The command line shows:

```
Command:_3drotate
Current positive angle in UCS:
  ANGDIR=counterclockwise ANGBASE=0
Select objects: pick the bowl 1 found
Select objects: right-click
Specify base point: pick the centre bottom of the
  bowl
Specify rotation angle or [Copy/Reference] <0>:
  enter 60 right-click
Command
```

Fig. 14.20 Example – 3D Rotate

2. Place in the **3D Navigate/SW Isometric** view and in **Conceptual** shading.

The result is shown in Fig. 14.20.

The Slice tool

First example – Slice (Fig. 14.24)

1. Construct a 3D model of the rod link device shown in the two-view projection (Fig. 14.21) on a layer colour green.

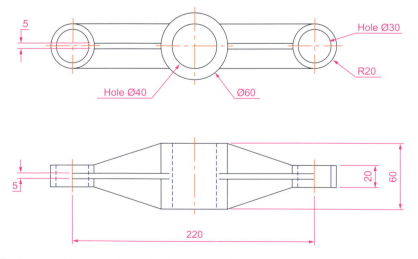

Fig. 14.21 First example – **Slice** – the two-view drawing

2. Place the 3D model in the **3D Navigation/Top** view.
3. *Call* the **Slice** tool from the **Home/Solid Editing** panel (Fig. 14.22).

Fig. 14.22 The **Slice** tool icon from the **Home/Solid Editing** panel

The command line shows:

```
Command:_slice
Select objects: pick the 3D model
Select objects to slice: right-click
Specify start point of slicing plane or [planar
  Object/Surface/Zaxis/View/XY/YZ/ZX/3points]
  <3points>: pick
Specify second point on plane: pick
Specify a point on desired side or [keep Both
  sides] <Both>: right-click
Command:
```

Fig. 14.23 shows the *picked* points.

start point   second point

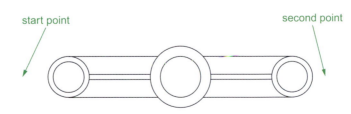

Fig. 14.23 First example – **Slice** – the *pick* points

Fig. 14.24 First
example – **Slice**

4. With the **Move** tool, move the lower half of the sliced model away from the upper half.
5. Place the 3D model(s) in the **ViewCube/Isometric** view.
6. Shade in **Conceptual** visual style. The result is shown in Fig. 14.24.

Second example – Slice (Fig. 14.25)

1. On a layer of colour **Green**, construct the closed pline shown in the left-hand drawing (Fig. 14.25) and with the **Revolve** tool, form a solid of revolution from the pline.
2. With the **Slice** tool and working to the same sequence as for the first **Slice** example, form two halves of the 3D model.
3. Place in **View/Views/Visual Styles/X-Ray**.

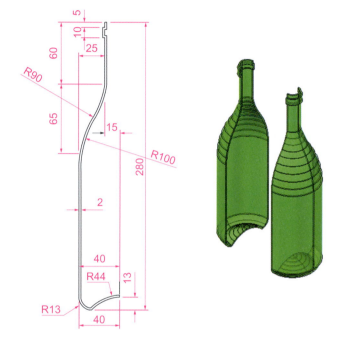

Fig. 14.25 Second example – **Slice**

The right-hand illustration of Fig. 14.25 shows the result.

4. Place the model in the **3D Navigate/Front** view, **Zoom** to **1** and **Move** its parts apart.
5. Make a new layer **Hatch** of colour **Magenta** and make the layer current.

Views of 3D models

Some of the possible viewing positions of a 3D model which can be obtained by using the **View/Views 3D Navigation** popup list have already been shown in earlier pages. Fig. 14.27 shows the viewing positions of the 3D model of the arrow (Fig. 14.26) using the viewing positions from the **3D Navigation** popup.

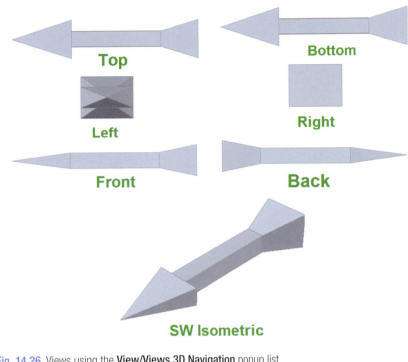

Fig. 14.26 Views using the **View/Views 3D Navigation** popup list

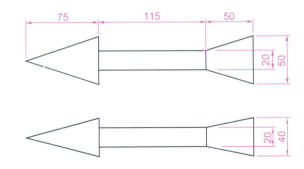

Fig. 14.27 Two views of the arrow

The ViewCube

Another method of obtaining viewing positions of a 3D model is by using the **ViewCube**, which can usually be seen at the top-right corner of the AutoCAD 2011 window (Fig. 14.28). The **ViewCube** can be turned off by *entering* **navvcubedisplay** at the command line and *entering* **1** as a response as follows:

```
Command: navvcubedisplay
Enter new value for NAVVCUBEDISPLAY <3>: enter 1
  right-click
```

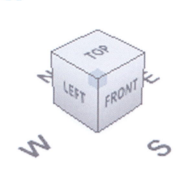

Fig. 14.28 The ViewCube

Entering **3** as a response to **navvcubedisplay** causes the **ViewCube** to reappear.

The **ViewCube** is used as follows:

Click on **Top** and the **Top** view of a 3D model appears.

Click on **Front** and the **Front** view of a 3D model appears.

And so on. *Clicking* the arrows at top, bottom or sides of the **ViewCube** moves a model between views.

A *click* on the house icon at the top of the **ViewCube** places a model in the **SW Isometric** view.

Using Dynamic Input to construct a helix

As with all other tools (commands) in AutoCAD 2011 a helix can be formed working with the **Dynamic Input** (DYN) system. Fig. 14.30 shows the stages (**1** to **5**) in the construction of the helix in the second example.

Set **DYN** on with a *click* on its button in the status bar.

1. *Click* the **Helix** tool icon in the **Home/Draw** panel (Fig. 14.29). The first of the **DYN** prompts appears. *Enter* the following at the command line using the down key of the keyboard when necessary.

```
Command: _Helix
Number of turns=10 Twist=CCW
Specify center point of base: enter 95,210
Specify base radius or [Diameter]: enter 55
```

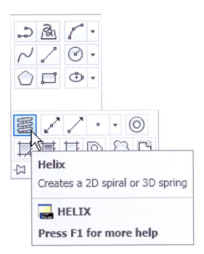

Fig. 14.29 The **Helix** tool in the **Home/Draw** panel

```
Specify top radius or [Diameter]: enter 35
Specify helix height or [Axis endpoint/Turns/turn
  Height/tWist]: enter 100
Command:
```

Fig. 14.30 shows the sequence of **DYN** tooltips and the completed helix.

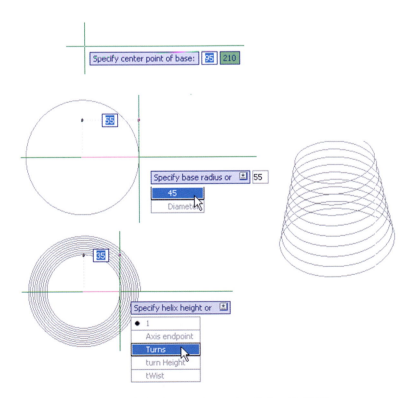

Fig. 14.30 Constructing the helix for the second example with the aid of **DYN**

3D Surfaces

As mentioned on page 245 surfaces can be formed using the **Extrude** tool on lines and polylines. Two examples are given below in Figs 14.39 and 14.41.

First example – 3D Surface (Fig. 14.39)

1. In the **ViewCube/Top** view, on a layer colour **Magenta**, construct the polyline (Fig. 14.31).

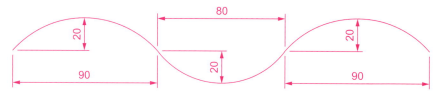

Fig. 14.31 First example – **3D Surface** – polyline to be extruded

2. In the **ViewCube/Isometric** view, call the **Extrude** tool from the **Home/Modeling** control and extrude the polyline to a height of **80**. The result is shown in Fig. 14.32.

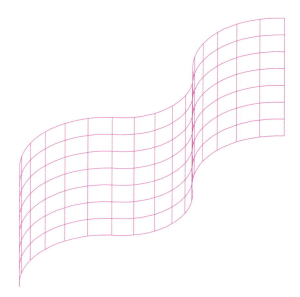

Fig. 14.32 First example – **3D Surface**

Second example – 3D Surface (Fig. 14.41)

1. In the **Top** view on a layer colour **Blue** construct the circle (Fig. 14.33) using the **Break** tool break the circle as shown.

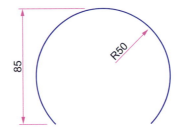

Fig. 14.33 Second example – **3D Surface**. The part circle to be extruded

2. In the **3D Manager/SW Isometric** view, call the **Extrude** tool and extrude the part circle to a height of **80**. Shade in the **Conceptual** visual style (Fig. 14.34).

The result is shown in Fig. 14.34.

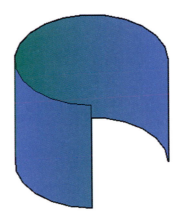

Fig. 14.34 Second example – **3D Surface**

REVISION NOTES

1. 3D models can be saved as blocks in a similar manner to the method of saving 2D drawings as blocks.
2. Libraries can be made up from 3D model drawings.
3. 3D models saved as blocks can be inserted into other drawings via the DesignCenter.
4. Arrays of 3D model drawings can be constructed in 3D space using the 3D Array tool.
5. 3D models can be mirrored in 3D space using the 3D Mirror tool.
6. 3D models can be rotated in 3D space using the 3D Rotate tool.
7. 3D models can be cut into parts with the Slice tool.
8. Helices can be constructed using the Helix tool.
9. Both the View/View/Navigation popup list and the ViewCube can be used for placing 3D models in different viewing positions in 3D space.
10. The Dynamic Input (DYN) method of construction can be used equally as well when constructing 3D model drawings as when constructing 2D drawings.
11. 3D surfaces can be formed from polylines or lines with Extrude.

Exercises

Methods of constructing answers to the following exercises can be found in the free website:

http://books.elsevier.com/companions/978-0-08-096575-8

1. Fig. 14.35 shows a **Realistic** shaded view of the 3D model for this exercise. Fig. 14.36 is a three-view projection of the model. Working to the details given in Fig. 14.36, construct the 3D model.

Fig. 14.35 Exercise 1 – a three-view projection

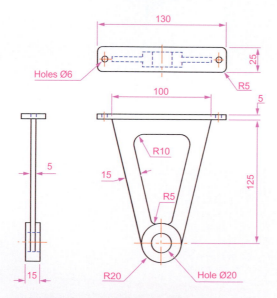

Fig. 14.36 Exercise 1 – a three-view projection

2. Construct a 3D model drawing of the separating link shown in the two-view projection (Fig. 14.37). With the **Slice** tool, slice the model into two parts and remove the rear part.

Place the front half in an isometric view using the **ViewCube** and shade the resulting model.

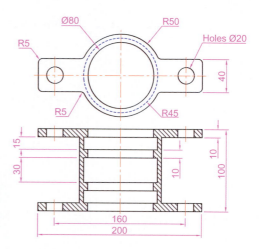

Fig. 14.37 Exercise 2

3. Working to the dimensions given in the two orthographic projections (Fig. 14.38), and working on two layers of different colours, construct an assembled 3D model of the one part inside the other.

With the **Slice** tool, slice the resulting 3D model into two equal parts, place in an isometric view. Shade the resulting model in **Realistic** mode as shown in Fig. 14.39.

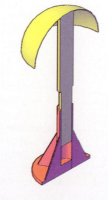

Fig. 14.39 Exercise 3

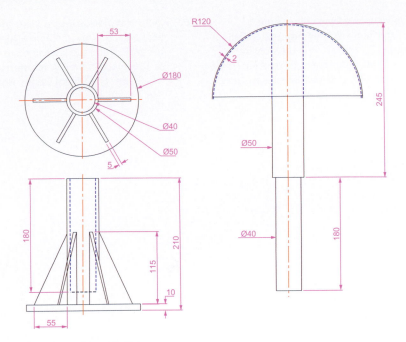

Fig. 14.38 Exercise 3 – orthographic projections

4. Construct a solid of revolution of the jug shown in the orthographic projection (Fig. 14.40). Construct a handle from an extrusion of a circle along a semicircular path. Union the two parts. Place the 3D model in a suitable isometric view and render.

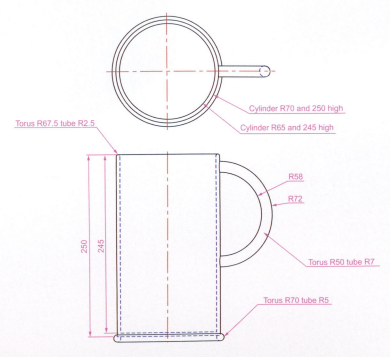

Fig. 14.40 Exercise 4

5. In the **Top** view on a layer colour **blue** construct the four polylines (Fig. 14.41). Call the **Extrude** tool and extrude the polylines to a height of **80** and place in the **Isometric** view. Then call **Visual Styles/Shades of Gray** shading (Fig. 14.42).

6. In **3D Navigation/Right** view construct the lines and arc (Fig. 14.43) on a layer colour **green**. Extrude the lines and arc to a height of **180**, place in the **SW Isometric** view and in the shade style **Visual Styles/Realistic** (Fig. 14.44).

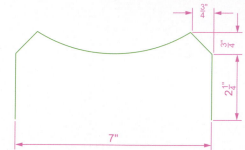

Fig. 14.41 Exercise 5 – outline to be extruded

Fig. 14.43 Exercise 6 – outline to be extruded

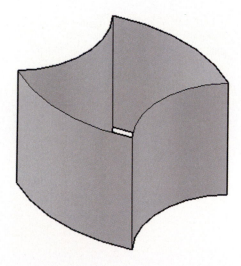

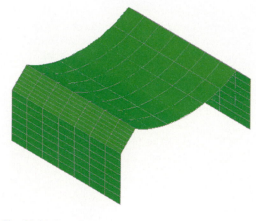

Fig. 14.44 Exercise 6

Fig. 14.42 Exercise 5

Rendering

AIMS OF THIS CHAPTER

The aims of this chapter are:

1. To construct a template for 3D Modeling to be used as the drawing window for further work in 3D in this book.
2. To introduce the use of the **Render** tools in producing photographic like images of 3D solid models.
3. To show how to illuminate a 3D solid model to obtain good lighting effects when rendering.
4. To give examples of the rendering of 3D solid models.
5. To introduce the idea of adding materials to 3D solid models in order to obtain a realistic appearance to a rendering.
6. To demonstrate the use of the forms of shading available using **Visual Styles** shading.
7. To demonstrate methods of printing rendered 3D solid models.
8. To give an example of the use of a camera.

Setting up a new 3D template

In this chapter we will be constructing all 3D model drawings in the **acadiso3D.dwt** template. The template is based on the **3D Modeling** workspace shown on page 258 in Chapter 13.

1. *Click* the **Workspace Switching** button and *click* **3D Modeling** from the menu which appears (Fig. 15.1).

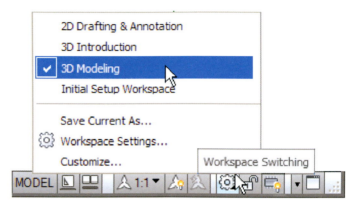

Fig. 15.1 *Click* **3D Modeling** in the **Workspace Settings menu**

2. The AutoCAD window (Fig. 15.2) appears.

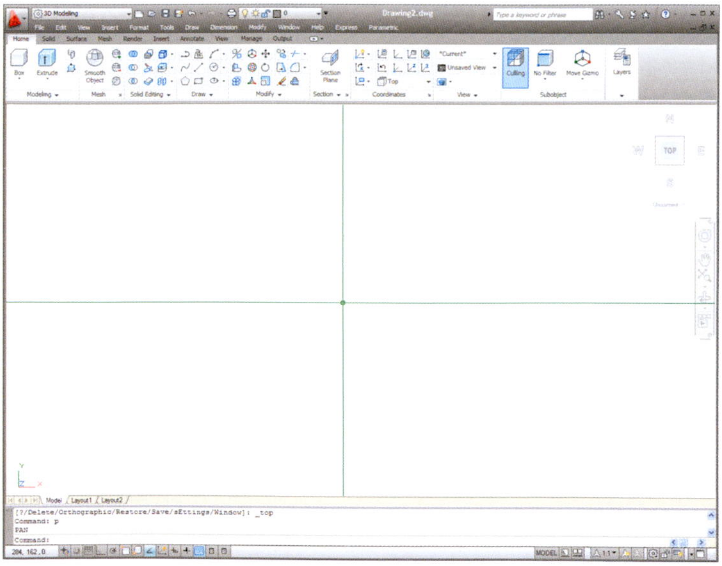

Fig. 15.2 The **3D Modeling** workspace

3. Set **Units** to a **Precision** of **0**, **Snap** to **5** and **Grid** to **10**. Set **Limits** to **420,297**. **Zoom** to **All**.

4. In the **Options** dialog *click* the **Files** tab and *click* **Default Template File Name for QNEW** followed by a *double-click* on the file name which appears. This brings up the **Select Template** dialog, from which the **acadiso3d.dwt** can be selected. Now when AutoCAD 2011 is opened from the Windows desktop, the acadiso3D.dwt template will open.

5. Set up five layers of different colours named after the colours.

6. Save the template to the name **acadiso3D** and then *enter* a suitable description in the **Template Definition** dialog.

The Materials Browser palette

Click **Materials Browser** in the **Render/Materials** palette (Fig. 15.3). The **Materials Browser** palette appears *docked* at an edge of the AutoCAD window. *Drag* the palette away from its *docked* position. *Click* the arrow to the left of **Autodesk Library** and in the list which appears, *click* **Brick**. A list of brick icons appears in a list to the right of the **Autodesk Library** list (Fig. 15.4).

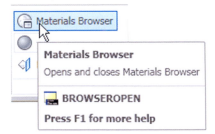

Fig. 15.3 The **Materials Browser** button in the **Render/Materials** panel

The **Materials Browser** palette can be *docked* against either side of the AutoCAD window if needed.

Applying materials to a model

Materials can be applied to a 3D model from selection of the icons in the **Materials Browser** palette. Three examples follow – applying a **Brick** material, applying a **Metal** material and applying a **Wood** material.

Fig. 15.4 The **Materials Browser** palette showing the **Brick** list

In the three examples which follow lighting effects are obtained by turning **Sun Status** on, by *clicking* the **Sun Status** icon in the **Render/Sun & Location** panel (Fig. 15.5). The command line shows:

```
Command: _sunstatus
Enter new value for SUNSTATUS <1>: 0
Command:
```

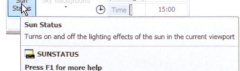

Fig. 15.5 The **Sun Status** button in the **Render/Sun % Location** panel

When the material has been applied, *click* **Render Region** from the sub-panel of the **Render/Render** panel (Fig. 15.6) and after selecting a window surrounding the model, the model renders (Fig. 15.6).

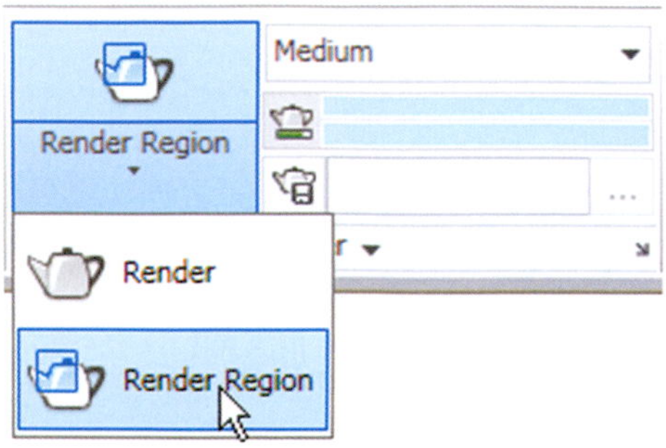

Fig. 15.6 The **Render Region** button from the **Render/Render** panel

First example – applying a **Masonry Brick** material (Fig. 15.7)

Construct the necessary 3D model (Fig. 15.8). In the **Material Browser** palette, in the **Autodesk Library** list *click* **Brick**. A number of icons appear in the right-hand column of the palette representing different brick types. *Pick* the **Brown Modular** icon from the list. The icon appears in the **Materials in this document** area of the palette. *Right-click* in the icon and in the menu which appears select **Assign to Selection**. *Click* the model. Select **Render Region** from the **Render/Render** panel (Fig. 15.6). Window the model. The model renders (Fig. 15.7).

Second example – applying a **Metal** material (Fig. 15.8)

Construct the necessary 3D model. From the **Materials Browser** palette *click* **Metals** in the **Autodesk Library** list. Select **polished Brass 7** from the metal icons. *Click* **Assign to Selection** from the *right-click* menu in the **Materials in this document** area and *click* the model. Then with the **Render Region** tool render the model (Fig. 15.8).

Third example – applying a **Wood** material (Fig. 15.9)

Construct the necessary 3D model – a board. In the **Materials Browser** palette *click* **Wood** in the **Autodesk Library** list. Select **Pine Coarse** from the wood icons . *Click* **Assign to Selection** from the *right-click* menu in the **Materials in this document** area and *click* the model. Then with the **Render Region** tool render the model (Fig. 15.9).

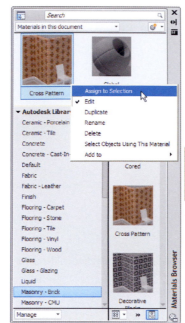

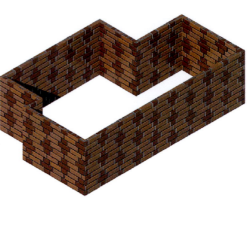

Fig. 15.7 First example – **assigning a Masonry Brick material**

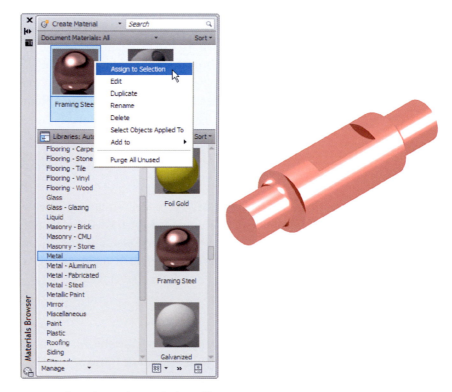

Fig. 15.8 Second example – **assigning a Metal material**

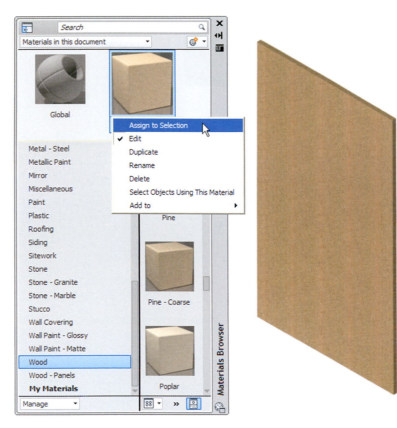

Fig. 15.9 Third example – **assigning a Wood material**

Modifying an applied material

If the result of applying a material direct to a model from the selected materials palette is not satisfactory, modifications to the applied material can be made. In the case of the third example, *double-click* on the chosen material icon in the **Materials Browser** palette and the **Materials Editor** palette appears showing the materials in the drawing (Fig. 15.10). Features such as colour of the applied material choosing different texture maps of the material (or materials) applied to a model can be amended as wished from this palette. In this example:

1. *Click* the arrow to the right of the **Image** area of the palette and a popup menu appears. Select **Wood** from this menu and the **Texture Editor** palette appears showing the material in its **Wood** appearance. In this palette a number of material changes can be made.

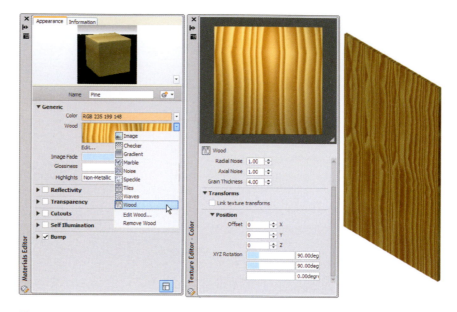

Fig. 15.10 The **Materials Browser** palette showing the materials in a 3D model and the **material Editor Open File** dialog

2. In this third example changes have been made to **Radial Noise**, **Axial Noise**, **Grain Thickness** and **XYZ Rotation**.
3. *Clicks* in the check boxes named **Reflectivity**, **Transparency**, etc. bring up features which can amend the material being edited.

Experimenting with this variety of settings in the **Materials Editor** palette allows emending the material to be used to the operator's satisfaction.

Note:
Material bitmaps are kept in the folders

C:\Program Files\Common Files\Autodesk\Shared\Materials 2001\ asset library fbm.\1\Mats (or 2\Mats or 3\Mats).

Fourth example – Available Materials in Drawing (Fig. 15.11)

As an example Fig. 15.11 shows the five of the materials applied to various parts of a 3D model of a hut in a set of fields surrounded by fences. The **Materials Browser** is shown. A *click* on a material in the **Available Materials in Drawing** brings the **Materials Editor** palette to screen, in which changes can be made to the selected material.

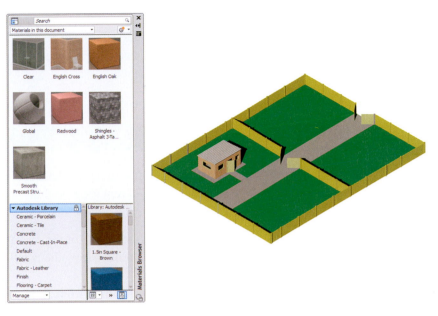

Fig. 15.11 An example of materials applied to parts of a 3D model

The Render tools and dialogs

The tool icons and menus in the **Render/Render** sub-panel are shown in Fig. 15.12.

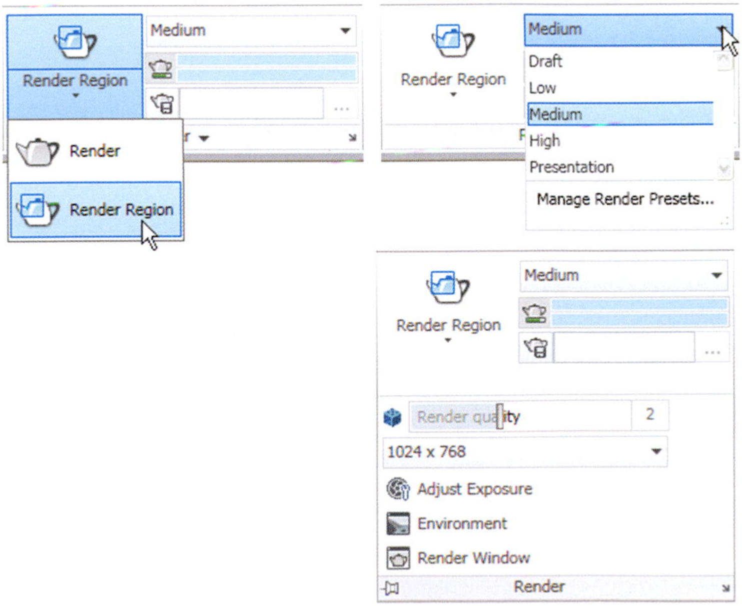

Fig. 15.12 The tools and menus in the **Render/Render** panel

A *click* in the outward facing arrow at the bottom right-hand corner of the **Render/Render** panel brings the **Advanced Render Settings** palette on screen. Note that a *click* on this arrow if it appears in any panel will bring either a palette or a dialog on screen.

The Lights tools

The different forms of lighting from light palettes are shown in Fig. 15.13. There are a large number of different types of lighting available when using AutoCAD 2011, among which those most frequently used are:

Default lighting. Depends on the setting of the set variable.

Point lights shed light in all directions from the position in which the light is placed.

Distant lights send parallel rays of light from their position in the direction chosen by the operator.

Spotlights illuminate as if from a spotlight. The light is in a direction set by the operator and is in the form of a cone, with a 'hotspot' cone giving a brighter spot on the model being lit.

Sun light can be edited as to position.

Sky background and illumination.

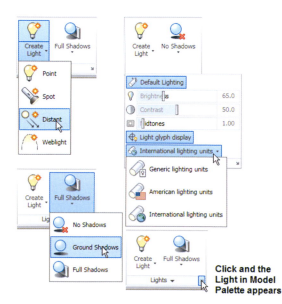

Fig. 15.13 **Lighting** buttons and menus in the **Render/Lights** panel

A variety of lights of different types in which lights of a selected wattage which can be placed in a lighting scene are available from the **Tool Palettes - All Palettes** palette. These are shown in Fig. 15.14.

The set variable **LIGHTINGUNITS** must be set to **1** or **2** for these lights to function. To set this variable:

```
Command: enter lightingunits right-click
Enter new value for LIGHTINGUNITS <2>:
```

Settings are:

0: No lighting units are used and standard (generic) lighting is enabled.
1: American lighting units (foot-candles) are used and photometric lighting is enabled.
2: International lighting units (lux) are used and photometric lighting is enabled.

Note: In the previous examples of rendering, **Generic lighting** was chosen.

Placing lights to illuminate a 3D model

In this book examples of lighting methods shown in examples will only be concerned with the use of **Point**, **Direct** and **Spot** lights, together with **Default lighting**, except for the example given on page 315, associated with using a camera.

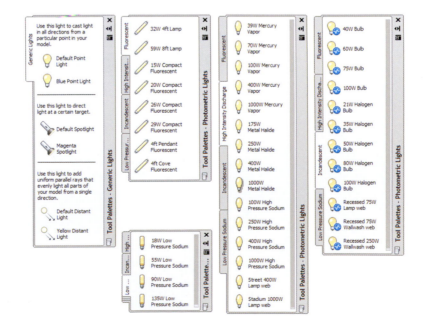

Fig. 15.14 The Lighting tool palettes

Any number of the three types of lights – **Point**, **Distant** and **Spotlight** – can be positioned in 3D space as wished by the operator.

In general, good lighting effects can be obtained by placing a **Point** light high above the object(s) being illuminated, with a **Distant** light placed pointing towards the object at a distance from the front and above the general height of the object(s) and with a second **Distant** light pointing towards the object(s) from one side and not as high as the first **Distant** light. If desired **Spotlights** can be used either on their own or in conjunction with the other two forms of lighting.

Setting rendering background colour

The default background colour for rendering in the acadiso3D template is black by default. In this book, all renderings are shown on a white background in the viewport in which the 3D model drawing was constructed. To set the background to white for renderings:

1. At the command line:

Command: *enter view right-click*

The **View Manager** dialog appears (Fig. 15.15). *Click* **Model View** in its **Views** list, followed by a *click* on the **New...** button.

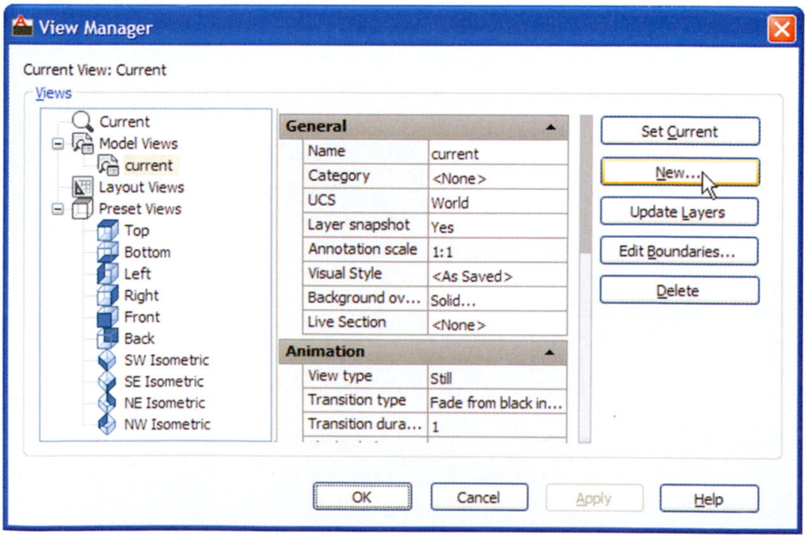

Fig. 15.15 The **View Manager** dialog

2. The **New View/Shot Properties** dialog (Fig. 15.16) appears. *Enter* **current** (or similar) in the **View name** field. In the **Background** popup list *click* **Solid**. The **Background** dialog appears (Fig. 15.17).

3. In the **Background** dialog *click* in the **Color** field. The **Select Color** dialog appears (Fig. 15.18).

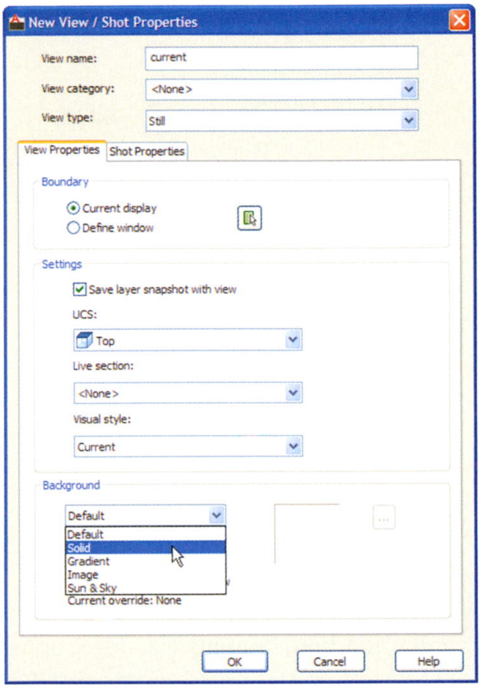

Fig. 15.16 The **New View/Shot Properties** dialog

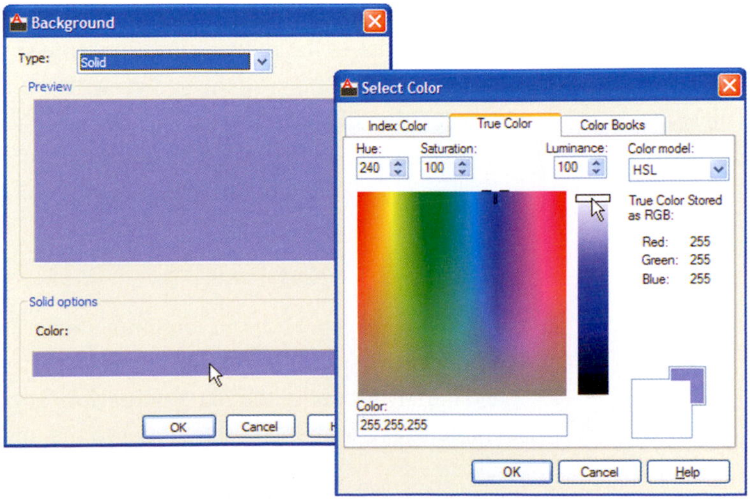

Fig. 15.17 The **Background** dialog

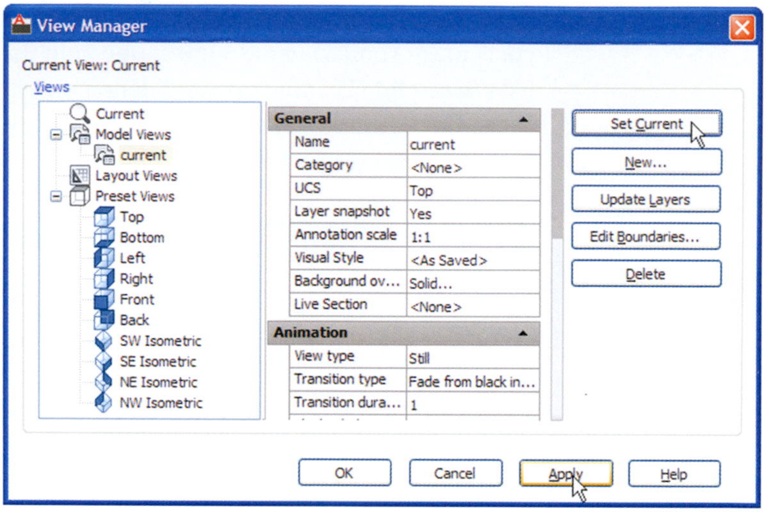

Fig. 15.18 The **View Manager** dialog

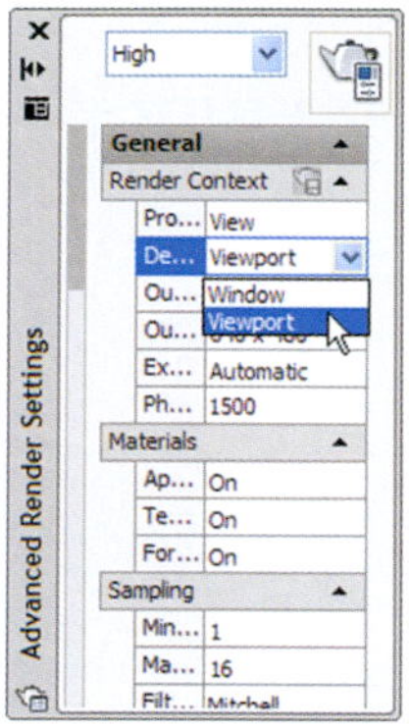

Fig. 15.19 The **Advanced Render Settings** dialog

4. In the **Select Color** dialog *drag* the slider as far upwards as possible to change the colour to white (**255,255,255**). Then *click* the dialog's **OK** button. The **Background** dialog reappears showing white in the **Color** and **Preview** fields. *Click* the **Background** dialog's **OK** button.

5. The **New View/Shot Properties** dialog reappears showing **current** highlighted in the **Views** list. *Click* the dialog's **OK** button.

6. The **View Manager** dialog reappears. *Click* the **Set Current** button, followed by a *click* on the dialog's **OK** button (Fig. 15.18).

7. *Enter* **rpref** at the command line. The **Advanced Render Settings** palette appears. In the palette, in the **Render Context** field *click* the arrow to the right of **Window** and in the popup menu which appears *click* **Viewport** as the rendering destination (Fig. 15.19).

8. Close the palette and save the screen with the new settings as the template **3dacadiso.dwt**. This will ensure renderings are made in the workspace in which the 3D model was constructed to be the same workspace in which renderings are made – on a white background.

First example – Rendering (Fig. 15.28)

1. Construct a 3D model of the wing nut shown in the two-view projection (Fig. 15.20).

2. Place the 3D model in the **3D Navigation/Top** view, **Zoom** to **1** and with the **Move** tool, move the model to the upper part of the AutoCAD drawing area.

3. *Click* the **Point Light** tool icon in the **Render/Lights** panel (Fig. 15.21). The warning window (Fig. 15.22) appears. *Click* **Turn off Default Lighting** in the window.

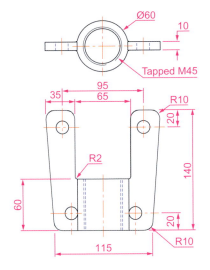

Fig. 15.20 First example – **Rendering** –
two-view projection

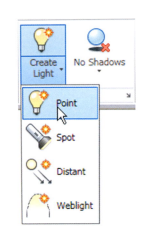

Fig. 15.21 The **Point Light** icon in the
Render/Lights panel

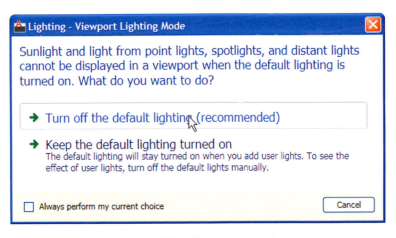

Fig. 15.22 The **Lighting – Viewport Lighting Mode** warning window

4. A **New Point Light** icon appears (depending upon the setting of the
   Light Glyph Setting in the **Drafting** area of the **Options** dialog) and
   the command line shows:

```
Command:_pointlight
Specify source location <0,0,0>: enter .xy
  right-click of pick centre of model (need Z):
  enter 500 right-click
Enter an option to change [Name/Intensity/Status/
  shadoW/Attenuation/Color/eXit]
<eXit>:enter n right-click
```

```
Enter light name <Pointlight1>: enter Point01
  right-click
Enter an option to change [Name/Intensity/
  Status/shadoW/Attenuation/Color/eXit]
  <eXit>: right-click
Command:
```

5. There are several methods by which **Distant** lights can be called. By selecting **Default Distant Light** from the **Generic Lights** palette (Fig. 15.29), with a *click* on the **Distant** icon in the **Render/Lights** panel, by *entering* **distantlight** at the command line.

No matter which method is adopted the **Lighting – Viewport Lighting Mode** dialog (Fig. 15.22) appears. *Click* **Turn off default lighting (recommended)**. The **Lighting - Photometric Distant Lights** dialog then appears (Fig. 15.23). *Click* **Allow distant lights** in this dialog and the command line shows:

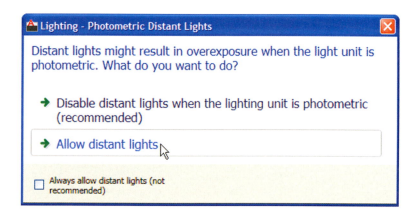

Fig. 15.23 The **Photometric Distant Lights** dialog

```
Command: _distantlight
Specify light direction FROM <0,0,0> or [Vector]:
  enter .xy right-click
of pick a point below and to the left of the model
  (need Z): enter 400 right-click
Specify light direction TO <1,1,1>: enter .xy
  right-click
of pick a point at the centre of the model (need Z):
  enter 70 right-click
Enter an option to change [Name/Intensity/Status/
  shadoW/Color/eXit] <eXit>: enter n right-click
Enter light name <Distantlight8>: enter Distant01
  right-click
```

```
Enter an option to change [Name/Intensity/Status/
  shadoW/Color/eXit] <eXit>: right-click
Command:
```

6. Place another **Distant Light** (**Distant2**) at the front and below the model **FROM Z** of **300** and at the same position **TO the model.**

7. When the model has been rendered if a light requires to be changed in intensity, shadow, position or colour, *click* the arrow at the bottom right-hand corner of the **Render/Lights** panel (Fig. 15.24) and the **Lights in Model** palette appears (Fig. 15.25). *Double-click* a light name in the palette and the **Properties** palette for the elected light appears into which modifications can be made (Fig. 15.25). Amendments can be made as thought necessary.

Fig. 15.24 The arrow at the bottom of the **Render/Lights** panel

> **Notes**
>
> 1. In this example the **Intensity factor** has been set at **0.5** for lights. This is possible because the lights are close to the model. In larger size models the **Intensity factor** may have to be set to a higher figure.
> 2. Before setting the **Intensity factor** to **0.5**, **Units** need setting to **OO** in the **Drawing Units** dialog (see Chapter 1).

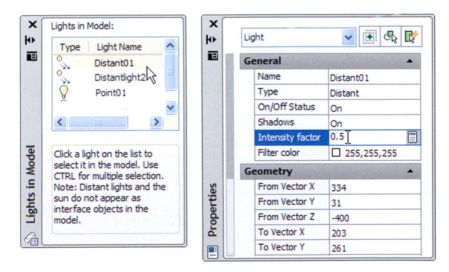

Fig. 15.25 The **Lights in Model** and **Properties** palettes

Assigning a material to the model

1. Open the **Materials Browser** palette, with a *click* on the **Materials Browser** icon in the **Render/Materials** panel. From the **Autodesk Library** list in the palette, select **Metals**. When the icons for the metals

appear in the right-hand column of the palette, *double-click* **Brass Polished**. The icon appears in the **Materials in this document** area of the palette (Fig. 15.26).

Fig. 15.26 The **Material Browser** and the rendering

2. *Click* **Assign to Selection** in the *right-click* menu of the material in the **Materials Browser** palette, followed by a *click* on the model, followed by a *left-click* when the model has received the assignment.
3. Select **Presentation** from the **Render Presets** menu in the sub **Render/ Render** panel (Fig. 15.27).

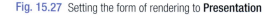

Fig. 15.27 Setting the form of rendering to **Presentation**

4. Render the model (Fig. 15.28) using the **Render Region** tool from the **Render/Render** panel and if now satisfied save to a suitable file name (Fig. 15.29).

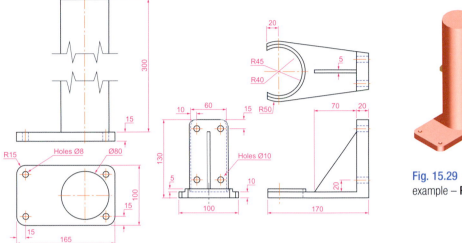

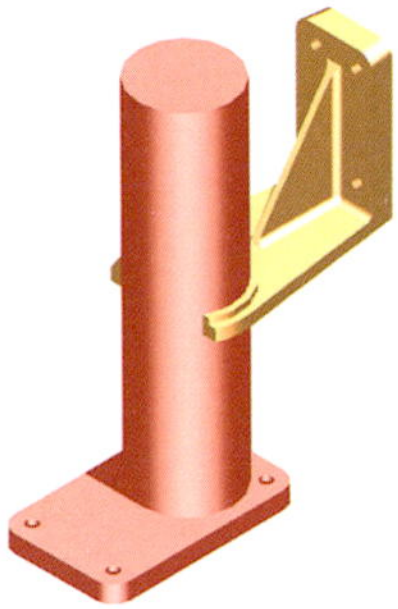

Fig. 15.29 Second example – **Rendering**

Fig. 15.28 Second example – **Rendering** – orthographic projection

---

Note

The limited descriptions of rendering given in these pages do not show the full value of different types of lights, materials and rendering methods. The reader is advised to experiment with the facilities available for rendering.

Second example – Rendering a 3D model (Fig. 15.29)

1. Construct 3D models of the two parts of the stand and support given in the projections (Fig. 15.28) with the two parts assembled together.
2. Place the scene in the **ViewCube/Top** view, **Zoom** to **1** and add lighting.
3. Add different materials to the parts of the assembly and render the result.

Fig. 15.28 shows the resulting rendering.

Third example – Rendering (Fig. 15.33)

Fig. 15.30 is an exploded, rendered 3D model of a pumping device from a machine and Fig. 15.31 is a third angle orthographic projection of the device.

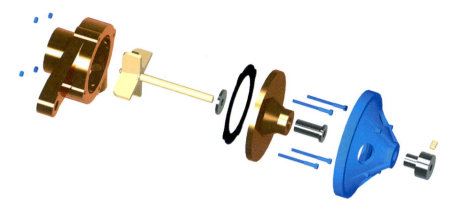

Fig. 15.30 Third example – **Rendering**

Free Orbit

Example – Free Orbit (Fig. 15.32)

Place the second example in a **Visual Styles/Conceptual** shading.

Click the **Free Orbit** button in the **View/Navigate** panel (Fig. 15.32). An orbit cursor appears on screen. Moving the cursor under mouse control allows the model on screen to be placed in any desired viewing position. Fig. 15.33 shows an example of a **Free Orbit**.

Right-click anywhere on screen and a right-click menu appears.

Producing hardcopy

Printing or plotting a drawing on screen from AutoCAD 2011 can be carried out from either **Model Space** or **Paper Space**.

First example – printing (Fig. 15.36)

This example is of a drawing which has been acted upon by the **Visual Styles/Realistic** shading mode.

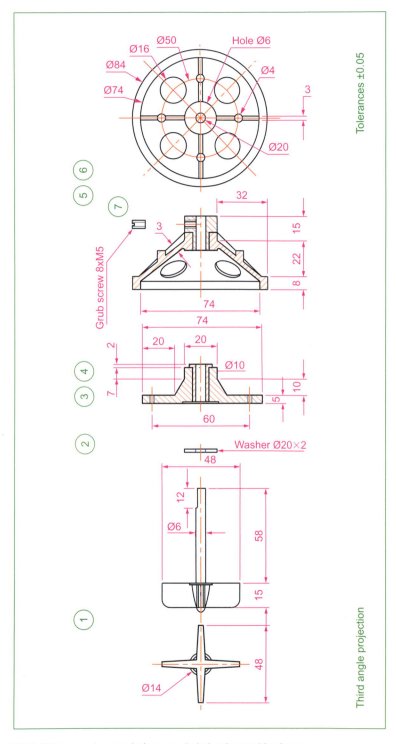

Fig. 15.31 Third example – **rendering – exploded orthographic views**

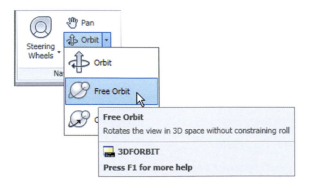

Fig. 15.32 The **Free Orbit** tool from the **View/Navigation** panel

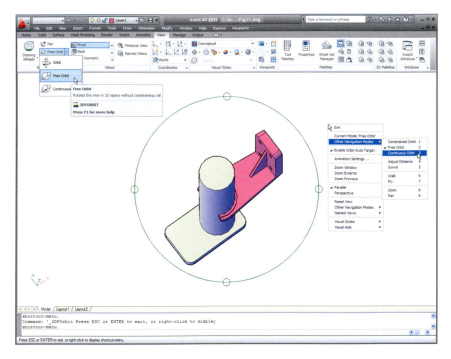

Fig. 15.33 Example – **Free Orbit**

1. With a drawing to be printed or plotted on screen *click* the **Plot** tool icon in the **Output/Plot** panel (Fig. 15.34).
2. The **Plot** dialog appears (Fig. 15.35). Set the **Printer/Plotter** to a printer or plotter currently attached to the computer and the **Paper Size** to a paper size to which the printer/plotter is set.
3. *Click* the **Preview** button of the dialog and if the preview is OK (Fig. 15.36), *right-click* and in the right-click menu which appears, *click* **Plot**. The drawing plots producing the necessary 'hardcopy'.

Fig. 15.34 The **Plot** icon in the **Output/Plot** panel

Fig. 15.35 The **Plot** dialog

Second example – multiple view copy (Fig. 15.37)

The 3D model to be printed is a **Realistic** view of a 3D model. To print a multiple view copy:

1. Place the drawing in a **Four: Equal** viewport setting.
2. Make a new layer **vports** of colour cyan and make it the current layer.

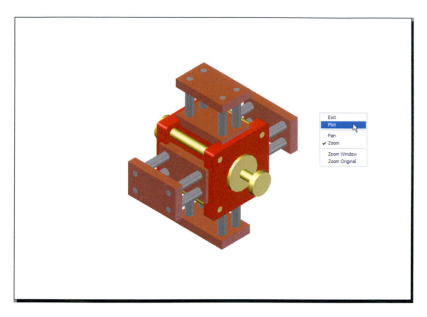

Fig. 15.36 First example – **Print Preview – printing a single copy**

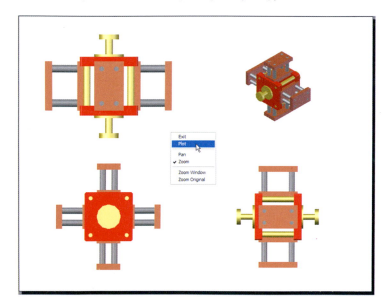

Fig. 15.37 Second example – **multiple view copy**

3. *Click* the **Layout** button in the status bar. At the command line:

```
Command: enter mv (MVIEW) right-click
MVIEW
Specify corner of viewport or [ON/OFF/Fit/
  Shadeplot/Lock/Object/Polygonal/Restore/
  LAyer/2/3/4] <Fit>: enter r (Restore)
right-click
```

```
Enter viewport configuration name or [?]
  <*Active>: right-click
Specify first corner or [Fit] <Fit>: right-click
Command:
```

The drawing appears in **Paper Space**. The views of the 3D model appear each within a cyan outline in each viewport.

4. Turn layer **vports** off. The cyan outlines of the viewports disappear.
5. *Click* the **Plot** tool icon in the **Output/Plot** toolbar. Make sure the correct **Printer/Plotter** and **Paper Size** settings are selected and *click* the **Preview** button of the dialog.
6. If the preview is satisfactory (Fig. 15.37), *right-click* and from the right-click menu *click* **Plot**. The drawing plots to produce the required four-viewport hardcopy.

Saving and opening 3D model drawings

3D model drawings are saved and/or opened in the same way as are 2D drawings. To save a drawing *click* **Save As…** in the **File** drop-down menu and save the drawing in the **Save Drawing As** dialog by *entering* a drawing file name in the **File Name** field of the dialog before *clicking* the **Save** button. To open a drawing which has been saved *click* **Open…** in the **File** drop-down menu, and in the **Select File** dialog which appears select a file name from the file list.

There are differences between saving a 2D and a 3D drawing, in that when 3D model drawing is shaded by using a visual style from the **Home/View** panel, the shading is saved with the drawing.

Camera

Example – Camera shot in room scene

This example is of a camera being used in a room in which several chairs, stools and tables have been placed. Start by constructing one of the chairs.

Constructing one of the chairs

1. In a **Top** view construct a polyline from an ellipse (after setting **pedit** to **1**), trimmed in half, then offset and formed into a single pline using **pedit**.
2. Construct a polyline from a similar ellipse, trimmed in half, then formed into a single pline using **pedit**.

3. Extrude both plines to suitable heights to form the chair frame and its cushion seat.
4. In a **Right** view, construct plines for the holes through the chair and extrude them to a suitable height and subtract them from the extrusion of the chair frame.
5. Add suitable materials and render the result (Fig. 15.38).

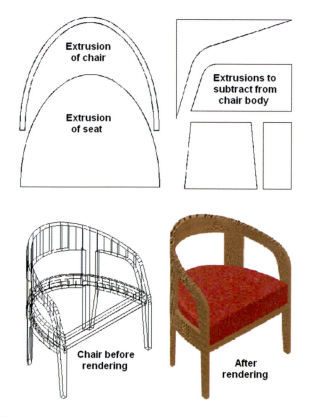

Fig. 15.38 Stages in constructing a chair

Constructing one of the stools

1. In the **Front** view and working to suitable sizes, construct a pline outline for one-quarter of the stool.
2. Extrude the pline to a suitable height.
3. **Mirror** the extrusion, followed by forming a union of the two mirrored parts.
4. In the **Top** view, copy the union, rotate the copy through 90 degrees, move it into a position across the original and form a union of the two.
5. Add a cylindrical cushion and render (Fig. 15.39).

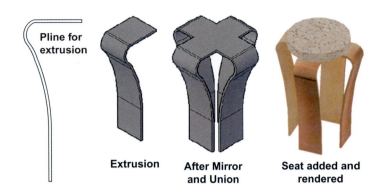

Fig. 15.39 Stages in constructing a stool

Constructing one of the tables

1. In the **Top** view and working to suitable sizes, construct a cylinder for the tabletop.
2. Construct two cylinders for the table rail and subtract the smaller from the larger.
3. Construct an ellipse from which a leg can be extruded and copy the extrusion 3 times to form the four legs.
4. In the **Front** view, move the parts to their correct positions relative to each other.
5. Add suitable materials and render (Fig. 15.40).

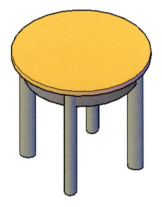

Fig. 15.40 A Conceptual shading of one of a table

Constructing walls, doors and window

Working to suitable sizes, construct walls, floor, doors and window using the **Box** tool (Fig. 15.41).

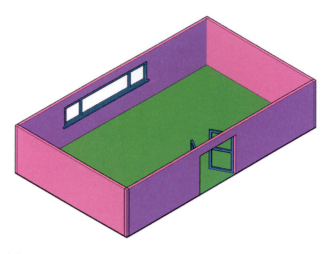

Fig. 15.41 A Conceptual style view of the walls, floor, doors and window

Using a camera

Inserting the furniture

In the **Top** view:

1. Insert the chair, copy it 3 times and move the copies to suitable positions.
2. Insert the stool, copy it 3 times and move the copies to suitable positions.
3. Insert the table, copy it 3 times and move the copies to suitable positions (Fig. 15.42).

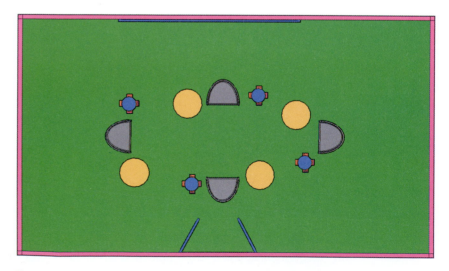

Fig. 15.42 **Top** view of the furniture inserted, copies and places in position

Adding lights

1. Place a **59 W 8 ft fluorescent** light central to the room just below the top of the wall height.
2. Place a **Point** light in the bottom right-hand central corner of the room (Fig. 15.43).

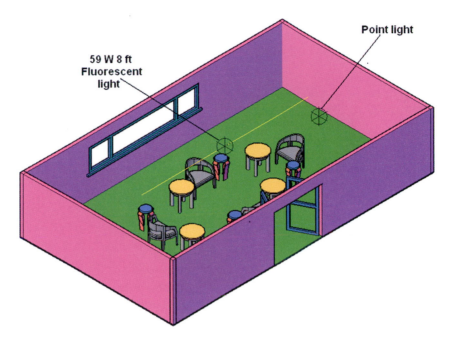

Point light

59 W 8 ft Fluorescent light

Fig. 15.43 Two lights placed in the room

Placing a camera

1. Place the scene in the **Front** view.
2. Select **Create Camera** from the **Render/Camera** panel or from the **View** drop-down menu (Fig. 15.44). The command line shows:

```
Command: _camera
Current camera settings: Height=0 Lens
  Length=80mm
Specify camera location: pick a position
Specify target location: drag to end of the cone
  into position
Enter an option [?/Name/LOcation/Height/Target/
  LEns/Clipping/View/eXit] <eXit>: enter
le (LEns) right-click
Specify lens length in mm <80>: enter 55
  right-click
```

```
Enter an option [?/Name/LOcation/Height/Target/
    LEns/Clipping/View/eXit] <eXit>: n
Enter name for new camera <Camera2>: right-click
    -accepts name (Camera1)
Enter an option [?/Name/LOcation/Height/Target/
    LEns/Clipping/View/eXit] <eXit>: right-click
Command:
```

And the camera will be seen in position (Fig. 15.45).

Fig. 15.44 Selecting **Create Camera** from the **View** drop-down menu

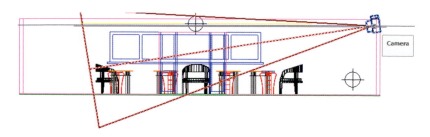

Fig. 15.45 The camera in position

3. At the command line *enter* **view**.

The **View Manager** dialog appears (Fig. 15.46). In the **Views** list *click* **Camera1**, followed by a *click* on the **Set Current** button, then the **OK** button. A view of the camera view fills the AutoCAD drawing area.

4. If not satisfied with the scene it can be amended in several ways from the **Camera/Swivel** command (**View** drop-down menu) and its *right-click* menu (Fig. 15.47).

The camera view (**Conceptual**) after amendment and before render is shown in Fig. 15.48.

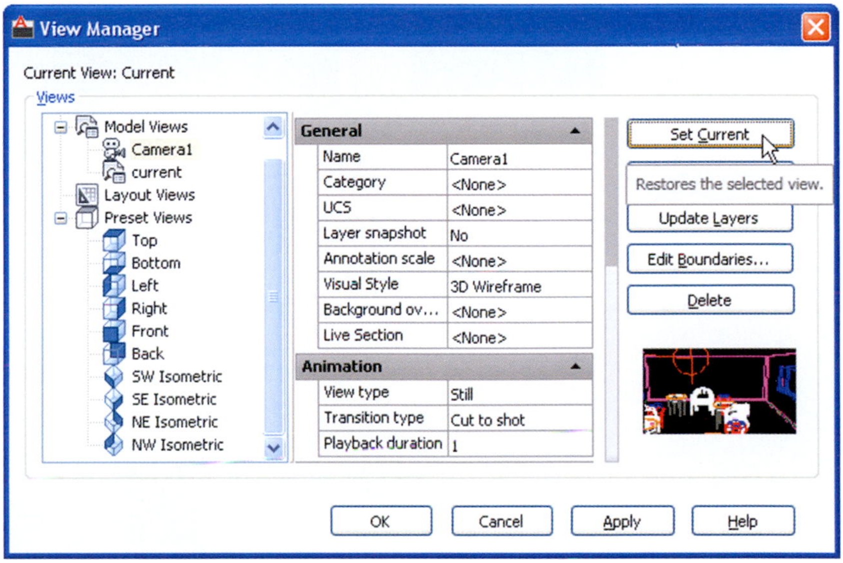

Fig. 15.46 Selecting **Camera1** from the **View Manager**

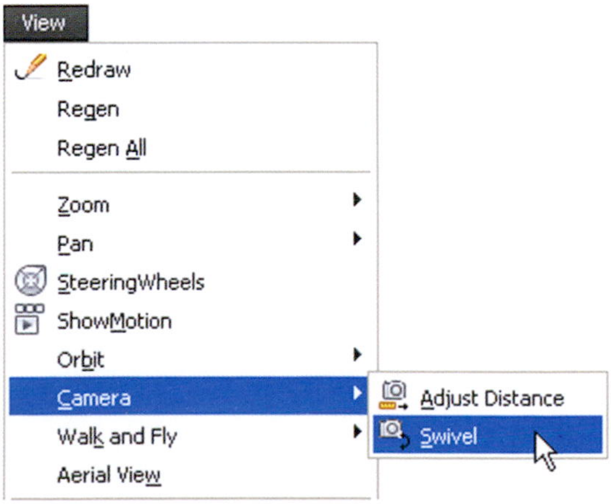

Fig. 15.47 Selecting **Camera/Swivel** from the **View** drop-down menu

CHAPTER 15

Fig. 15.48 The camera view (**Conceptual**) after amendment and before render

Other features of this scene

1. A fair number of materials were attached to objects as shown in the **Materials Browser** palette associated with the scene (Fig. 15.49).

Fig. 15.49 The materials in the scene as seen in the **Materials** palette

2. Changing the lens to different lens lengths can make appreciable differences to the scene. One rendering of the same room scene taken with a lens of **55 mm** is shown in Fig. 15.50 and another with a **100 mm** lens is shown in Fig. 15.51.

Fig. 15.50 The rendering of the scene taken with a **55 mm** lens

Fig. 15.51 The rendering of a scene taken with a **100 mm** lens camera

CHAPTER 15

REVISION NOTES

1. 3D models can be constructed in any of the workspaces – 2D Design & Annotation, 3D Basics or 3D Modeling. In Part 2 of this book 3D models are constructed in either the 3D Basics or the 3D Modeling workspace.
2. 3D model drawings can be constructed in either a Parallel projection or a Perspective projection layout.
3. Material and light palettes can be selected from the Render panels.
4. Materials can be modified from the Materials Editor palette.
5. In this book lighting of a scene with 3D models is mostly by placing two distant lights in front of and above the models, with one positioned to the left and the other to the right, and a point light above the centre of the scene. The exception is the lighting of the camera scenes on pages 315.
6. There are many other methods of lighting a scene, in particular using default lighting or sun lighting.
7. Several Render preset methods of rendering are available, from Draft to Presentation.
8. The use of the Orbit tools allows a 3D model to be presented in any position.
9. Plotting or printing of either Model or Layout windows is possible.
10. Hardcopy can be from a single viewport or from multiple viewports. When printing or plotting 3D model drawings Visual Style layouts print as they appear on screen.

Exercises

Methods of constructing answers to the following exercises can be found in the free website:

http://books.elsevier.com/companions/978-0-08-096575-8

1. A rendering of an assembled lathe tool holder is shown in Fig. 15.52. The rendering includes different materials for each part of the assembly.

    Working to the dimensions given in the parts orthographic drawing (Fig. 15.53), construct a 3D model drawing of the assembled lathe tool holder on several layers of different colours, add lighting and materials and render the model in an isometric view.

    Shade with **3D Visual Styles/Hidden** and print or plot a **ViewCube/Isometric** view of the model drawing.

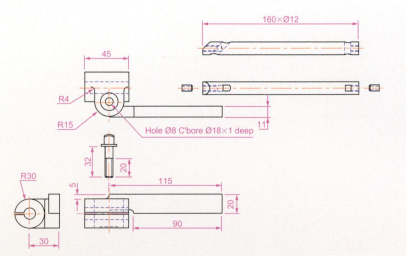

Fig. 15.52 Exercise 1

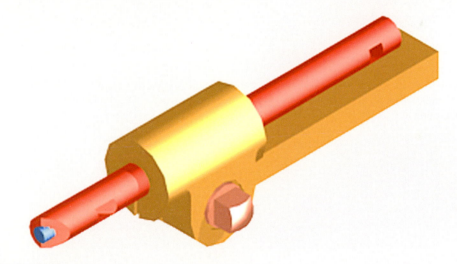

Fig. 15.53 Exercise 1 – parts drawings

2. Fig. 15.54 is a rendering of a drip tray. Working to the sizes given in Fig. 15.55, construct a 3D model drawing of the tray. Add lighting and a suitable material, place the model in an isometric view and render.

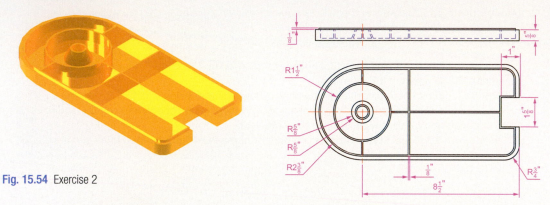

Fig. 15.54 Exercise 2

Fig. 15.55 Exercise 2 – two-view projection

3. A three-view drawing of a hanging spindle bearing in third angle orthographic projection is shown in Fig. 15.56. Working to the dimensions in the drawing construct a 3D model drawing of the bearing. Add lighting and a material and render the model.

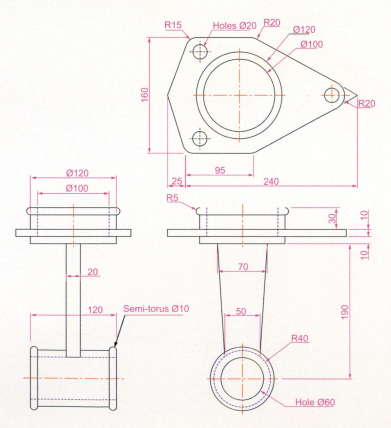

Fig. 15.56 Exercise 3

Building drawing

Building drawings

There are a number of different types of drawings related to the construction of any form of building. In this chapter a fairly typical example of a set of building drawings is shown. There are seven drawings related to the construction of an extension to an existing two-storey house (44 Ridgeway Road). These show:

1. A site plan of the original two-storey house, drawn to a scale of **1:200** (Fig. 16.1).

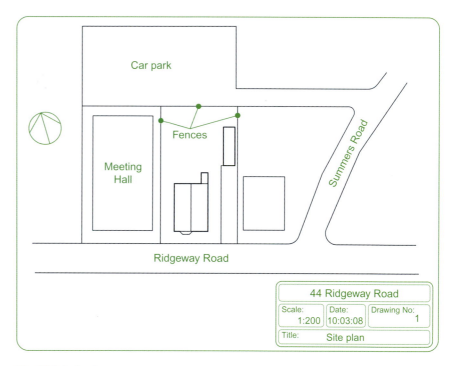

Fig. 16.1 A site plan

2. A site layout plan of the original house, drawn to a scale of **1:100** (Fig. 16.2).
3. Floor layouts of the original house, drawn to a scale of **1:50** (Fig. 16.3).
4. Views of all four sides of the original house, drawn to a scale of **1:50** (Fig. 16.4).
5. Floor layouts including the proposed extension, drawn to a scale of **1:50** (Fig. 16.5).
6. Views of all four sides of the house including the proposed extension, drawn to a scale of **1:50** (Fig. 16.6).
7. A sectional view through the proposed extension, drawn to a scale of **1:50** (Fig. 16.7).

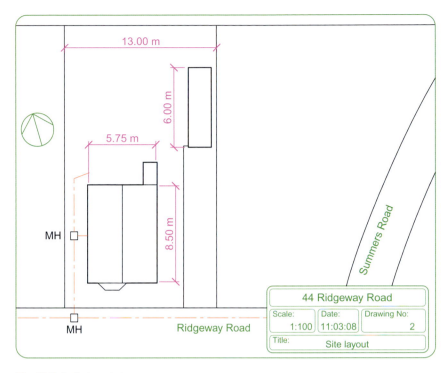

Fig. 16.2 A site layout plan

Fig. 16.3 Floor layouts drawing of the original house

Fig. 16.4 Views of the original house

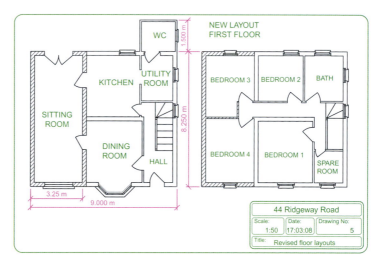

Fig. 16.5 Floor layouts drawing of the proposed extension

Notes

1. Other types of drawings will be constructed such as drawings showing the details of parts such as doors, windows and floor structures. These are often shown in sectional views.

2. Although the seven drawings related to the proposed extension of the house at 44 Ridgeway Road are shown here as having been constructed on either A3 or A4 layouts, it is common practice to include several types of building drawings on larger sheets such as A1 sheets of a size 820 mm by 594 mm.

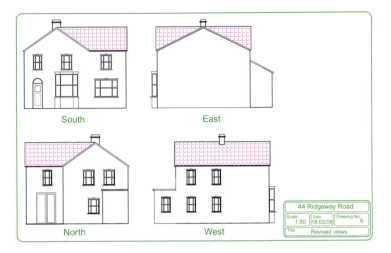

Fig. 16.6 Views including the proposed extension

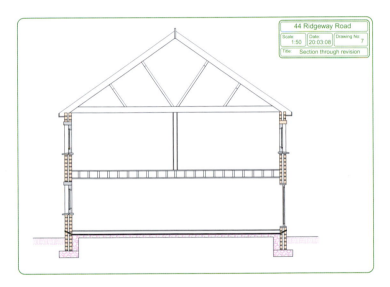

Fig. 16.7 A section through the proposed extension

Floor layouts

When constructing floor layout drawings it is advisable to build up a library of block drawings of symbols representing features such as doors and windows. These can then be inserted into layouts from the DesignCenter. A suggested small library of such block symbols is shown in Fig. 16.8.

Details of shapes and dimensions for the first two examples are taken from the drawings of the building and its extension at 44 Ridgeway Road given in Figs 16.2–16.6.

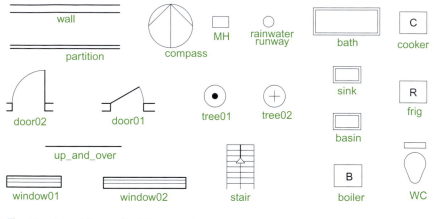

Fig. 16.8 A small library of building symbols

3D models of buildings

Details of the first example are taken from Figs 16.2–16.4 on pages 329 and 330.

The following steps describe the construction of a 3D model of 44 Ridgeway Road prior to the extension being added.

First example – 44 Ridgeway Road – original building

1. In the **Layer Properties Manager** palette – **Doors** (colour **red**), **Roof** (colour **green**), **Walls** (colour **blue**), **Windows** (colour **8**) (Fig. 16.9).
2. Set the screen to the **ViewCube/Front** view (Fig. 16.10).

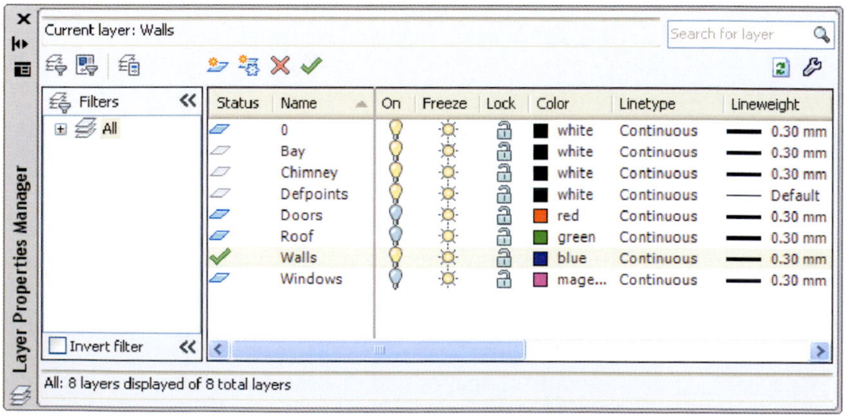

Fig. 16.10 Set screen to the **ViewCube/Front** view

Fig. 16.9 First example – the layers on which the model is to be constructed

3. Set the layer **Walls** current and, working to a scale of **1:50**, construct outlines of the walls. Construct outlines of the bay, windows and doors inside the wall outlines.
4. **Extrude** the wall, bay, window and door outlines to a height of **1**.
5. **Subtract** the bay, window and door outlines from the wall outlines. The result is shown in Fig. 16.11.

Fig. 16.11 First example – the walls

6. Make the layer **Windows** current and construct outlines of three of the windows which are of different sizes. Extrude the copings and cills to a height of **1.5** and the other parts to a height of **1**. Form a union of the main outline, the coping and the cill. The window pane extrusions will have to be subtracted from the union. Fig. 16.12 shows the 3D models of the three windows in a **ViewCube/Isometric** view.

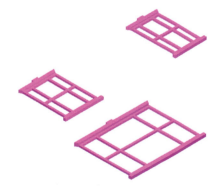

Fig. 16.12 First example – extrusions of the three sizes of windows

7. Move and copy the windows to their correct positions in the walls.
8. Make the layer **Doors** current and construct outlines of the doors and extrude to a height of **1**.

Fig. 16.13 First example – **Realistic** view of a 3D model of the chimney

9. Make layer **Chimney** current and construct a 3D model of the chimney (Fig. 16.13).
10. Make the layer **Roofs** current and construct outlines of the roofs (main building and garage) (see Fig. 16.14).

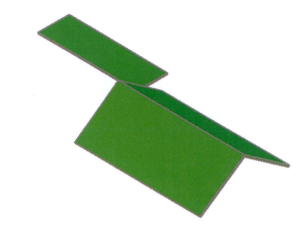

Fig. 16.14 First example – **Realistic** view of the roofs

11. On the layer **Bay** construct the bay and its windows.

Assembling the walls

Fig. 16.15 Set screen to **ViewCube/Top** view

1. Place the screen in the **ViewCube/Top** view (Fig. 16.15).
2. Make the layer **Walls** current and turn off all other layers other than **Windows**.
3. Place a window around each wall in turn. Move and/or rotate the walls until they are in their correct position relative to each other.
4. Place in the **ViewCube/Isometric** view and using the **Move** tool, move the walls into their correct positions relative to each other. Fig. 16.16 shows the walls in position in a **ViewCube/Top** view.

Fig. 16.16 First example – the four walls in their correct positions relative to each other in a **ViewCube/Top** view

5. Move the roof into position relative to the walls and move the chimney into position on the roof. Fig. 16.17 shows the resulting 3D model in a **ViewCube/Isometric** view (Fig. 16.18).

Fig. 16.18 Set screen to a **ViewCube/Isometric** view

Fig. 16.17 First example – a **Realistic** view of the assembled walls, windows, bay, roof and chimney

The garage

On layers **Walls** construct the walls and on layer **Windows** construct the windows. Fig. 16.19 is a **Realistic** visual style view of the 3D model as constructed so far.

Fig. 16.19 First example – **Realistic** view of the original house and garage

Second example – extension to 44 Ridgeway Road

Working to a scale of **1:50** and taking dimensions from the drawing Figs 16.5 and 16.6 and in a manner similar to the method of constructing the 3D model of the original building, add the extension to the original building. Fig. 16.20 shows a **Realistic** visual style view of the resulting 3D model. In this 3D model floors have been added – a ground and a first storey floor constructed on a new layer **Floors** of colour yellow. Note the changes in the bay and front door.

Fig. 16.20 Second example – a **Realistic** view of the building with its extension

Third example – small building in fields

Working to a scale of **1:50** from the dimensions given in Fig. 16.21, construct a 3D model of the hut following the steps given below.

The walls are painted concrete and the roof is corrugated iron.

In the **Layer Properties Manager** dialog make the new levels as follows:

Walls – colour **Blue**
Road – colour **Red**
Roof – colour **Red**
Windows – **Magenta**
Fence – colour **8**
Field – colour **Green**

Following the methods used in the construction of the house in the first example, construct the walls, roof, windows and door of the small building in one of the fields. Fig. 16.22 shows a **Realistic** visual style view of a 3D model of the hut.

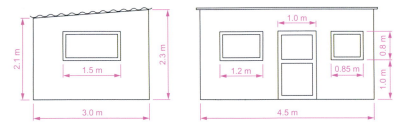

Fig. 16.21 Third example – front and end views of the hut

Fig. 16.22 Third example – a **Realistic** view of a 3D model of the hut

Constructing the fence, fields and road

1. Place the screen in a **Four: Equal** viewports setting.
2. Make the **Garden** layer current and in the **Top** viewport, construct an outline of the boundaries to the fields and to the building. Extrude the outline to a height of **0.5**.
3. Make the **Road** layer current and in the **Top** viewport, construct an outline of the road and extrude the outline to a height of **0.5**.
4. In the **Front** view, construct a single plank and a post of a fence and copy them a sufficient number of times to surround the four fields leaving gaps for the gates. With the **Union** tool form a union of all the posts and planks. Fig. 16.23 shows a part of the resulting fence in a **Realistic** visual style view in the **Isometric** viewport. With the **Union** tool form a union of all the planks and posts in the entire fence.
5. While still in the layer **Fence**, construct gates to the fields.
6. Make the **Road** layer current and construct an outline of the road. Extrude to a height of **0.5**.

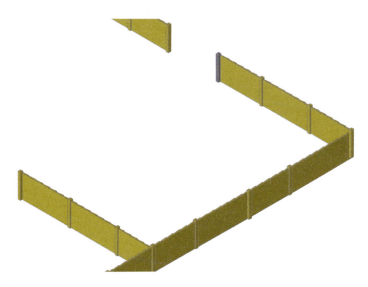

Fig. 16.23 Third example – part of the fence

Note

When constructing each of these features it is advisable to turn off those layers on which other features have been constructed.

Fig. 16.24 shows a **Conceptual** view of the hut in the fields with the road, fence and gates.

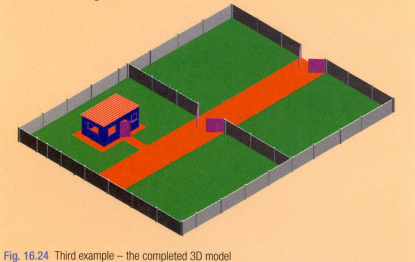

Fig. 16.24 Third example – the completed 3D model

Completing the second example

Working in a manner similar to the method used when constructing the roads, garden and fences for the third example, add the paths, garden area and fences and gates to the building 44 Ridgeway Road with its extension. Fig. 16.24 is a **Conceptual** visual style view of the resulting 3D model.

Material attachments and rendering

Second example

The following materials were attached to the various parts of the 3D model (Fig. 16.25). To attach the materials, all layers except the layer on which the objects to which the attachment of a particular material is being made are tuned off, allowing the material in question to be attached only to the elements to which each material is to be attached.

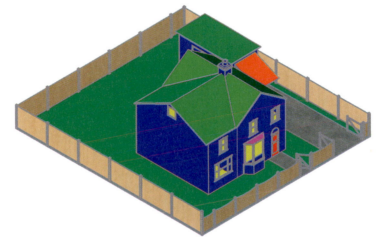

Fig. 16.25 Second example – the completed 3D model

Default: colour **7**
Doors: Wood Hickory
Fences: Wood – Spruce
Floors: Wood – Hickory
Garden: Green
Gates: Wood – White
Roofs: Brick – Herringbone
Windows: Wood – White

The 3D model was then rendered with **Output Size** set to **1024 × 768** and **Render Preset** set to **Presentation**, with **Sun Status** turned on. The resulting rendering is shown in Fig. 16.26.

Third example

Fig. 16.27 shows the third example after attaching materials and rendering.

CHAPTER 16

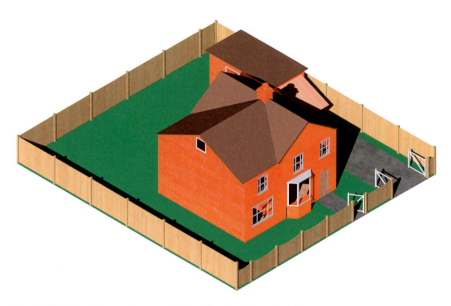

Fig. 16.26 Second example – a rendering after attaching materials

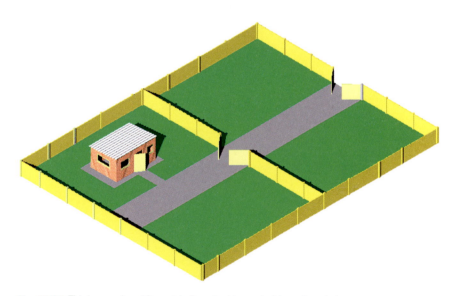

Fig. 16.27 Third example – 3D model after attaching materials and rendering

REVISION NOTES

1. There are a number of different types of building drawings – site plans, site layout plans, floor layouts, views, sectional views, detail drawings. AutoCAD 2011 is a suitable CAD program to use when constructing building drawings.

2. AutoCAD 2011 is a suitable CAD program for the construction of 3D models of buildings.

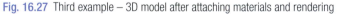

Exercises

Methods of constructing answers to the following exercises can be found in the free website:
http://books.elsevier.com/companions/978-0-08-096575-8

1. Fig. 16.28 is a site plan drawn to a scale of 1:200 showing a bungalow to be built in the garden of an existing bungalow. Construct the library of symbols shown in Fig. 16.8 on page 332 and by inserting the symbols from the DesignCenter construct a scale 1:50 drawing of the floor layout plan of the proposed bungalow.

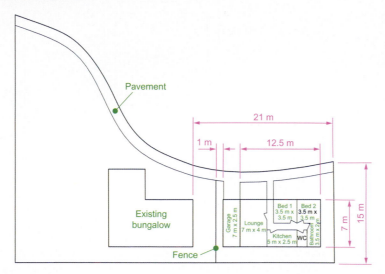

Fig. 16.28 Exercise 1

2. Fig. 16.29 is a site plan of a two-storey house of a building plot. Design and construct to a scale 1:50, a suggested pair of floor layouts for the two floors of the proposed house.

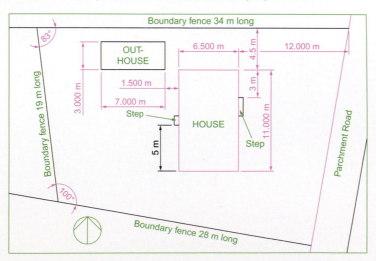

Fig. 16.29 Exercise 2

3. Fig. 16.30 shows a scale 1:100 site plan for the proposed bungalow 4 Caretaker Road. Construct the floor layout for the proposed house shown in Fig. 16.28.

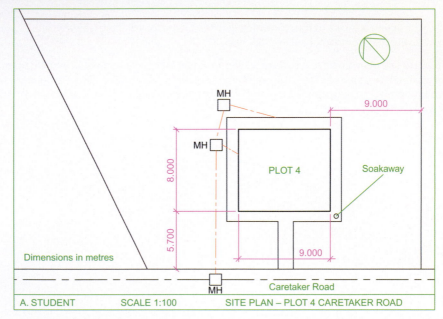

Fig. 16.30 Exercise 3 – site plan

4. Fig. 16.31 shows a building plan of a house in the site plan (Fig. 16.30). Construct a 3D model view of the house making an assumption as to the roofing and the heights connected with your model.

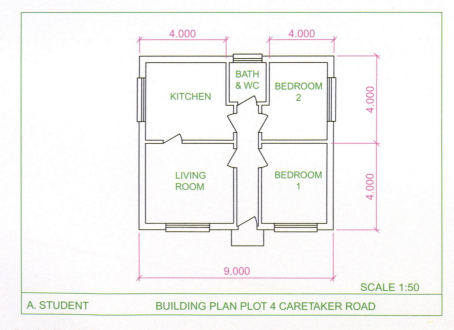

Fig. 16.31 Exercise 3 – a building

5. Fig. 16.32 is a three-view, dimensioned orthographic projection of a house. Fig. 16.33 is a rendering of a 3D model of the house. Construct the 3D model to a scale of 1:50, making estimates of dimensions not given in Fig. 16.32 and render using suitable materials.

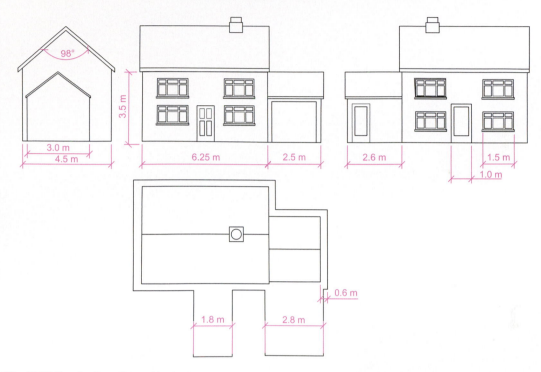

Fig. 16.32 Exercise 5 – orthographic views

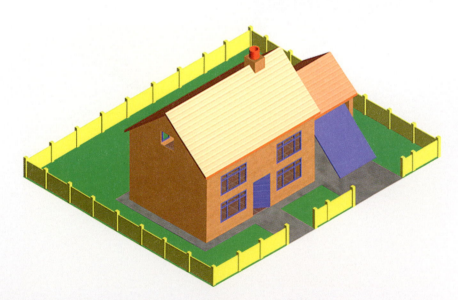

Fig. 16.33 Exercise 5 – the rendered model

6. Fig. 16.34 is a two-view orthographic projection of a small garage. Fig. 16.35 shows a rendering of a 3D model of the garage. Construct the 3D model of the garage working to a suitable scale.

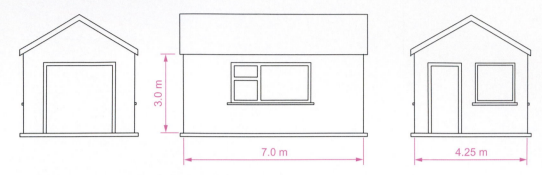

Fig. 16.34 Exercise 5 – orthographic views

Fig. 16.35 Exercise 5

Three-dimensional space

AIM OF THIS CHAPTER

The aim of this chapter is to show in examples the methods of manipulating 3D models in 3D space using tools – the **UCS** tools from the **View/Coordinates** panel or from the command line.

3D space

So far in this book, when constructing 3D model drawings, they have been constructed on the AutoCAD 2011 coordinate system which is based upon three planes:

The **XY Plane** – the screen of the computer.
The **XZ Plane** at right angles to the **XY Plane** and as if coming towards the operator of the computer.
A third plane (**YZ**) is lying at right angles to the other two planes (Fig. 17.1).

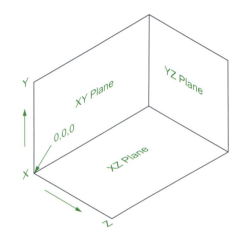

Fig. 17.1 The 3D space planes

In earlier chapters the **3D Navigate** drop-down menu and the **ViewCube** have been described to enable 3D objects which have been constructed on these three planes to be viewed from different viewing positions. Another method of placing the model in 3D space using the **Orbit** tool has also been described.

The User Coordinate System (UCS)

The **XY** plane is the basic **UCS** plane, which in terms of the ucs is known as the ***WORLD*** plane.

The **UCS** allows the operator to place the AutoCAD coordinate system in any position in 3D space using a variety of **UCS** tools (commands). Features of the **UCS** can be called either by *entering* **ucs** at the command line or by the selection of tools from the **View/Coordinates** panel (Fig. 17.2). Note

Fig. 17.2 The **View/ Coordinates** panel

Fig. 17.3 The drop-down menu from **World** in the panel

that a *click* on **World** in the panel brings a drop-down menu from which other views can be selected (Fig. 17.3).

If **ucs** is *entered* at the command line, it shows:

```
Command: enter ucs right-click
Current ucs name: *WORLD*
Specify origin of UCS or [Face/NAmed/OBject/
  Previous/View/World/X/Y/Z/ZAxis] <World>:
```

And from these prompts selection can be made.

The variable UCSFOLLOW

UCS planes can be set from using the methods shown in Figs 17.2 and 17.3 or by *entering* **ucs** at the command line. No matter which method is used, the variable **UCSFOLLOW** must first be set on as follows:

```
Command: enter ucsfollow right-click
Enter new value for UCSFOLLOW <0>: enter 1
  right-click
Command:
```

The UCS icon

The **UCS** icon indicates the directions in which the three coordinate axes **X**, **Y** and **Z** lie in the AutoCAD drawing. When working in 2D, only the

CHAPTER 17

X and **Y** axes are showing, but when the drawing area is in a 3D view all three coordinate arrows are showing, except when the model is in the **XY** plane. The icon can be turned off as follows:

```
Command: enter ucsicon right-click
Enter an option [ON/OFF/All/Noorigin/ORigin/
    Properties] <ON>:
```

To turn the icon off, *enter* **off** in response to the prompt line and the icon disappears from the screen.

The appearance of the icon can be changed by *entering* **p** (Properties) in response to the prompt line. The **UCS Icon** dialog appears in which changes can be made to the shape, line width and colour of the icon if wished.

Types of UCS icon

The shape of the icon can be varied partly when changes are made in the **UCS Icon** dialog but also according to whether the AutoCAD drawing area is in 2D, 3D or Paper Space (Fig. 17.4).

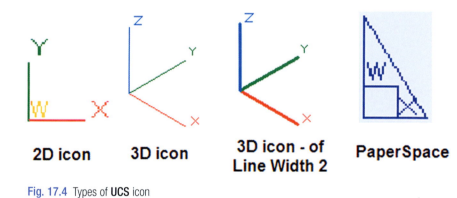

2D icon   **3D icon**   **3D icon - of Line Width 2**   **PaperSpace**

Fig. 17.4 Types of **UCS** icon

Examples of changing planes using the UCS

First example – changing UCS planes (Fig. 17.6)

1. Set **UCSFOLLOW** to **1** (ON).
2. Make a new layer colour **Red** and make the layer current. Place the screen in **ViewCube/Front** and **Zoom** to **1**.
3. Construct the pline outline (Fig. 17.5) and extrude to **120** high.
4. Place in **ViewCube/Isometric** view and **Zoom** to **1**.
5. With the **Fillet** tool, fillet corners to a radius of **20**.

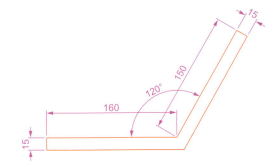

Fig. 17.5 First example – **Changing UCS planes** – pline for extrusion

6. At the command line:

```
Command: enter ucs right-click
Current ucs name: *WORLD*
Specify origin of UCS or [Face/NAmed/OBject/
  Previous/View/World/X/Y/Z/ZAxis] <World>:
  enter f (Face) right-click
Select face of solid object: pick the sloping
  face - its outline highlights
Enter an option [Next/Xflip/Yflip] <accept>:
  right-click
Regenerating model.
Command:
```

And the 3D model changes its plane so that the sloping face is now on the new UCS plane. **Zoom** to **1**.

7. On this new UCS, construct four cylinders of radius 7.5 and height – 15 (note the minus) and subtract them from the face.

8. *Enter* **ucs** at the command line again and *right-click* to place the model in the ***WORLD* UCS**.

9. Place four cylinders of the same radius and height into position in the base of the model and subtract them from the model.

10. Place the 3D model in a **ViewCube/Isometric** view and set in the **Home/View/Conceptual** visual style (Fig. 17.6).

Fig. 17.6 First example – **Changing UCS planes**

Second example – UCS (Fig. 17.9)

The 3D model for this example is a steam venting valve – a two-view third angle projection of the valve is shown in Fig. 17.7.

1. Make sure that **UCSFOLLOW** is set to **1**.

2. Place in the **UCS *WORLD*** view. Construct the **120** square plate at the base of the central portion of the valve. Construct five cylinders for the holes in the plate. Subtract the five cylinders from the base plate.

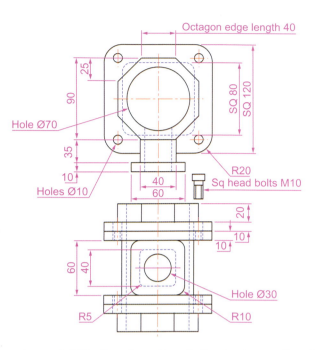

Fig. 17.7 Second example UCS – The orthographic projection of a steam venting valve

3. Construct the central part of the valve – a filleted **80** square extrusion with a central hole.

4. At the command line:

```
Command: enter ucs right-click
Current ucs name: *WORLD*
Specify origin of UCS or [Face/NAmed/OBject/
  Previous/View/World/X/Y/Z/ZAxis] <World>:
  enter x right-click
Specify rotation angle about X axis <90>:
  right-click
Command:
```

and the model assumes a **Front** view.

5. With the **Move** tool, move the central portion vertically up by **10**.

6. With the **Copy** tool, copy the base up to the top of the central portion.

7. With the **Union** tool, form a single 3D model of the three parts.

8. Make the layer **Construction** current.

9. Place the model in the **UCS *WORLD*** view. Construct the separate top part of the valve – a plate forming a union with a hexagonal plate and with holes matching those of the other parts.

10. Place the drawing in the **UCS X** view. Move the parts of the top into their correct positions relative to each other. With **Union** and **Subtract** complete the part. This will be made easier if the layer **0** is turned off.

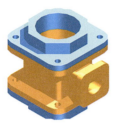

Fig. 17.8 **Second example UCS** – step **11** + rendering

11. Turn layer **0** back on and move the top into its correct position relative to the main part of the valve. Then with the **Mirror** tool, mirror the top to produce the bottom of the assembly (Fig. 17.8).

12. While in the **UCS X** view construct the three parts of a 3D model of the extrusion to the main body.

13. In the **UCS *WORLD*** view, move the parts into their correct position relative to each other. **Union** the two filleted rectangular extrusions and the main body. **Subtract** the cylinder from the whole (Fig. 17.9).

Fig. 17.9 **Second example UCS** – steps **12** and **13** + rendering

14. In the **UCS X** view, construct one of the bolts as shown in Fig. 17.10, forming a solid of revolution from a pline. Then construct a head to the bolt and with **Union** add it to the screw.

15. With the **Copy** tool, copy the bolt 7 times to give 8 bolts. With **Move**, and working in the **UCS *WORLD*** and **X** views, move the bolts into their correct positions relative to the 3D model.

16. Add suitable lighting and attach materials to all parts of the assembly and render the model.

17. Place the model in the **ViewCube/Isometric** view.

18. Save the model to a suitable file name.

19. Finally move all the parts away from each other to form an exploded view of the assembly (Fig. 17.11).

Fig. 17.10 **Second example UCS** – pline for the bolt

Third example – UCS (Fig. 17.15)

1. Set **UCSFOLLOW** to **1**.
2. Place the drawing area in the **UCS X** view.
3. Construct the outline (Fig 17.12) and extrude to a height of **120**.
4. *Click* the **3 Point** tool icon in the **View/Coordinates** panel (Fig. 17.13):

```
Command: _ucs
Current ucs name: *WORLD*
Specify origin of UCS or [Face/NAmed/OBject/
  Previous/View/World/X/Y/Z/ZAxis] <World>: _3
Specify new origin point <0,0,0>: pick point
  (Fig. 17.14)
Specify point on positive portion of X-axis: pick
  point (Fig. 17.14)
Specify point on positive-Y portion of the UCS XY
  plane <-142,200,0>: enter .xy right-click
of pick new origin point (Fig. 17.14) (need Z):
  enter 1 right-click
Regenerating model
Command:
```

Fig. 17.11 **Second example UCS**

CHAPTER 17

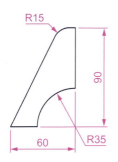

Fig. 17.12 Third example **UCS** – outline for 3D model

Fig. 17.14 shows the UCS points and the model regenerates in this new 3 point plane.

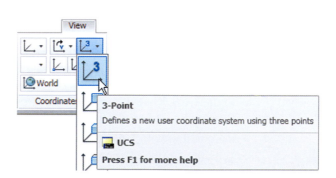

Fig. 17.13 The **UCS, 3 Point** icon in the **View/Coordinates** panel

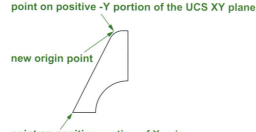

point on positive -Y portion of the UCS XY plane

new origin point

point on positive portion of X-axis

Fig. 17.14 Third example **UCS** – the three UCS points

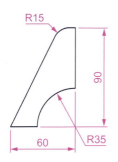

Fig. 17.15 Third example **UCS**

5. On the face of the model construct a rectangle **80 × 50** central to the face of the front of the model, fillet its corners to a radius of **10** and extrude to a height of **10**.
6. Place the model in the **ViewCube/Isometric** view and fillet the back edges of the second extrusion to a radius of **10**.
7. Subtract the second extrusion from the first.
8. Add lights and a suitable material, and render the model (Fig. 17.15).

Fourth example – UCS (Fig. 17.17)

1. With the last example still on screen, place the model in the **UCS *WORLD*** view.
2. Call the **Rotate** tool from the **Home/Modify** panel and rotate the model through 225 degrees.

3. *Click* the **X** tool icon in the **View/Coordinates** panel (Fig. 17.16):

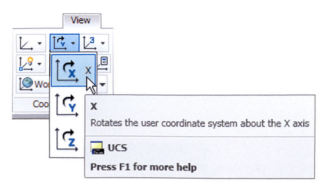

Fig. 17.16 The **UCS X** tool in the **View/Coordinates** panel

```
Command: _ucs
Current ucs name: *WORLD*
Specify origin of UCS or [Face/NAmed/OBject/
   Previous/View/World/X/Y/Z/ZAxis] <World>: _x
Specify rotation angle about X axis
   <90>: right-click
Regenerating model
Command:
```

Fig. 17.17 Fourth example

4. Render the model in its new **UCS** plane (Fig. 17.17).

Saving UCS views

If a number of different **UCS** planes are used in connection with the construction of a 3D model, each view obtained can be saved to a different name and recalled when required. To save a UCS plane view in which a 3D model drawing is being constructed *enter* **ucs** at the command line:

```
Current ucs name: *NO NAME*
Specify origin of UCS or [Face/NAmed/OBject/
   Previous/View/World/X/Y/Z/ZAxis] <World>:
   enter s right-click
Enter name to save current UCS or [?]: enter New
   View right-click
Regenerating model
Command:
```

Click the **UCS Settings** arrow in the **View/Coordinates** panel and the **UCS** dialog appears. *Click* the **Named UCSs** tab of the dialog and the names of views saved in the drawing appear (Fig. 17.18).

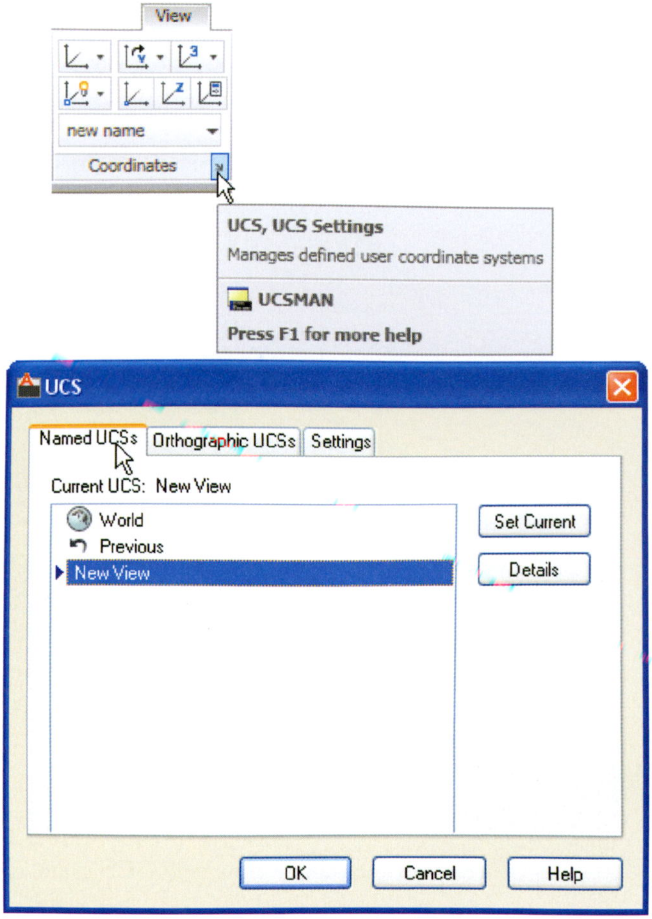

Fig. 17.18 The **UCS** dialog

Constructing 2D objects in 3D space

In previous chapters, there have been examples of 2D objects constructed with the **Polyline**, **Line**, **Circle** and other 2D tools to form the outlines for extrusions and solids of revolution. These outlines have been drawn on planes in the **ViewCube** settings.

First example – 2D outlines in 3D space (Fig. 17.21)

1. Construct a **3point UCS** to the following points:

```
Origin point: 80,90
X-axis point: 290,150
Positive-Y point: .xy of 80,90
(need Z): enter 1
```

2. On this **3point UCS** construct a 2D drawing of the plate to the dimensions given in Fig. 17.19, using the **Polyline**, **Ellipse** and **Circle** tools.
3. Save the **UCS** plane in the **UCS** dialog to the name **3point**.
4. Place the drawing area in the **ViewCube/Isometric** view (Fig. 17.20).
5. Make the layer **Red** current.
6. With the **Region** tool form regions of the 6 parts of the drawing and with the **Subtract** tool, subtract the circles and ellipse from the main outline.
7. Place in the **View/Visual Style/Realistic** visual style. Extrude the region to a height of **10** (Fig. 17.21).

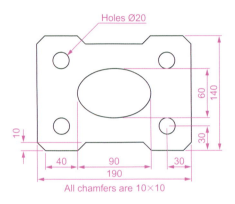

Fig. 17.19 First example – **2D outlines in 3D space**

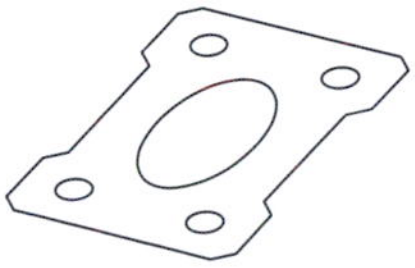

Fig. 17.20 First example – **2D outlines in 3D space**. The outline in the Isometric view

Fig. 17.21 First example – **2D outlines in 3D space**

Second example – 2D outlines in 3D space (Fig. 17.25)

1. Place the drawing area in the **ViewCube/Front** view, **Zoom** to **1** and construct the outline (Fig. 17.22).
2. Extrude the outline to **150** high.
3. Place in the **ViewCube/Isometric** view and **Zoom** to **1**.

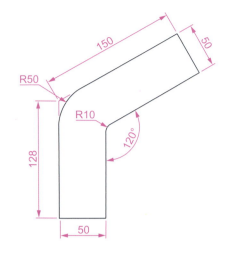

Fig. 17.22 Second example – **2D outlines in 3D space**. Outline to be extruded

4. *Click* the **Face** tool icon in the **View/Coordinates** panel (Fig. 17.23) and place the 3D model in the ucs plane shown in Fig. 17.24, selecting the sloping face of the extrusion for the plane and again **Zoom** to **1**.
5. With the **Circle** tool draw five circles as shown in Fig. 17.24.
6. Form a region from the five circles and with **Union** form a union of the regions.
7. Extrude the region to a height of **−60** (note the minus) – higher than the width of the sloping part of the 3D model.
8. Place the model in the **ViewCube/Isometric** view and subtract the extruded region from the model.
9. With the **Fillet** tool, fillet the upper corners of the slope of the main extrusion to a radius of **30**.

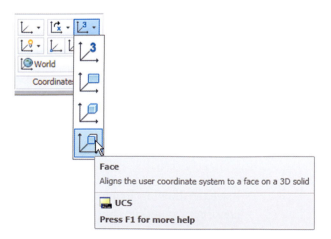

Fig. 17.23 The **Face** icon from the **View/Coordinates** panel

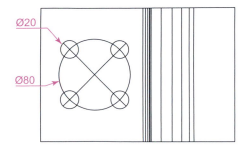

Fig. 17.24 Second example – **2D outlines in 3D space**

Fig. 17.25 Second example – **2D outlines in 3D space**

10. Place the model into another **UCS FACE** plane and construct a filleted pline of sides **80** and **50** and filleted to a radius of **20**. Extrude to a height of **-60** and subtract the extrusion from the 3D model.

11. Place in the **ViewCube/Isometric** view, add lighting and a material.

The result is shown in Fig. 17.25.

The Surfaces tools

The construction of 3D surfaces from lines, arc and plines has been dealt with – see pages 245 to 247 and 286 to 287. In this chapter examples of 3D surfaces constructed with the tools **Edgesurf**, **Rulesurf** and **Tabsurf** will be described. The tools can be called from the **Mesh Modeling/Primitives** panel. Fig. 17.26 shows the **Tabulated Surface** tool icon in the panel. The two icons to the right of that shown are the **Ruled Surface** and the **Edge Surface** tools. In this chapter these three surface tools will be called by *entering* their tool names at the command line.

Fig. 17.26 The **Tabulated Surface** tool icon in the **Mesh Modeling/Primitives**

CHAPTER 17

Surface meshes

Surface meshes are controlled by the set variables **Surftab1** and **Surftab2**. These variables are set as follows:

At the command line:

```
Command: enter surftab1 right-click
Enter new value for SURFTAB1 <6>: enter 24
  right-click
Command:
```

The Edgesurf tool – Fig. 17.29

1. Make a new layer colour **magenta**. Make that layer current.
2. Place the drawing area in the **View Cube/Right** view. **Zoom** to **All**.
3. Construct the polyline to the sizes and shape as shown in Fig. 17.27.
4. Place the drawing area in the **View Cube/Top** view. **Zoom** to **All**.
5. Copy the pline to the right by **250**.
6. Place the drawing in the **ViewCube/Isometric** view. **Zoom** to **All**.
7. With the **Line** tool, draw lines between the ends of the two plines using the **endpoint** osnap (Fig. 17.28). Note that if polylines are drawn they will not be accurate at this stage.
8. Set **SURFTAB1** to **32** and **SURFTAB2** to **64**.
9. At the command line:

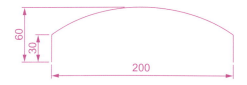

Fig. 17.27 Example – **Edgesurf** – pline outline

Fig. 17.28 Example – Edgesurf – adding lines joining the plines

```
Command: enter edgesurf right-click
Current wire frame density: SURFTAB1=32
  SURFTAB2=64
Select object 1 for surface edge: pick one of the
  lines (or plines)
Select object 2 for surface edge: pick the next
  adjacent line (or pline)
Select object 3 for surface edge: pick the next
  adjacent line (or pline)
Select object 4 for surface edge: pick the last
  line (or pline)
Command:
```

The result is shown in Fig. 17.29.

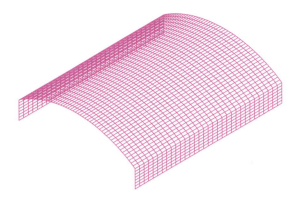

Fig. 17.29 Example – **Edgesurf**

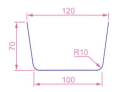

Fig. 17.30 **Rulesurf** – the outline

The *Rulesurf* tool – Fig. 17.29

1. Make a new layer colour **blue** and make the layer current.
2. In the **ViewCube/Front** view construct the pline as shown in Fig. 17.30.
3. In the **3D Navigate/Top**, **Zoom** to **1** and copy the pline to a vertical distance of **120**.
4. Place in the **3D Navigate/Southwest Isometric** view and **Zoom** to **1**
5. Set **SURFTAB1** to **32**.
6. At the command line:

```
Command: enter rulesurf right-click
Current wire frame density: SURFTAB1=32
Select first defining curve: pick one of the plines
Select second defining curve: pick the other pline
Command:
```

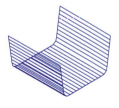

Fig. 17.31 Example – **Rulesurf**

The result is given in Fig. 17.31.

The *Tabsurf* tool – Fig. 17.32

1. Make a new layer of colour **red** and make the layer current.
2. Set **Surftab1** to **2**.
3. In the **ViewCube/Top** view construct a hexagon of edge length **35**.
4. In the **ViewCube/Front** view and in the centre of the hexagon construct a pline of height **100**.
5. Place the drawing in the **ViewCube/Isometric** view.
6. At the command line:

```
Command: enter tabsurf right-click
Current wire frame density: SURFTAB1=2
```

```
Select objects for path curve: pick the hexagon
Select object for direction vector: pick the pline
Command:
```

See Fig. 17.32.

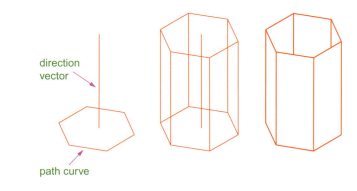

Fig. 17.32 Example – **Tabsurf**

REVISION NOTES

1. The UCS tools can be called from the **View/Coordinates** panel or by entering ucs at the command line.
2. The variable **UCSFOLLOW** must first be set ON (to **1**) before operations of the UCS can be brought into action.
3. There are several types of UCS icon – 2D, 3D and Pspace.
4. The position of the plane in 3D space on which a drawing is being constructed can be varied using tools from the **View/Coordinates** panel.
5. The planes on which drawings constructed on different planes in 3D space can be saved in the **UCS** dialog.
6. The tools **Edgesurf**, **Rulesurf** and **Tabsurf** can be used to construct surfaces in addition to surfaces which can be constructed from plines and lines using the **Extrude** tool.

Exercises

Methods of constructing answers to the following exercises can be found in the free website:
http://books.elsevier.com/companions/978-0-08-096575-8

1. Fig. 17.33 is a rendering of a two-view projection of an angle bracket in which two pins are placed in holes in each of the arms of the bracket. Fig. 17.34 is a two-view projection of the bracket.

    Construct a 3D model of the bracket and its pins.

    Add lighting to the scene and materials to the parts of the model and render.

2. The two-view projection (Fig. 17.35) shows a stand consisting of two hexagonal prisms. Circular holes have been cut right through each face of the smaller hexagonal prism and rectangular holes with rounded ends have been cut right through the faces of the larger.

    Construct a 3D model of the stand. When completed add suitable lighting to the scene. Then add a material to the model and render (Fig. 17.36).

Fig. 17.33 Exercise 1 – a rendering

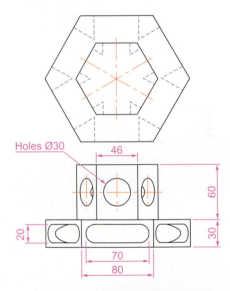

Fig. 17.35 Exercise 2 – details of shapes and sizes

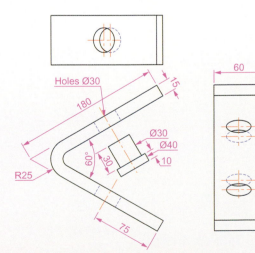

Fig. 17.34 Exercise 1 – details of shape and sizes

Fig. 17.36 Exercise 2 – a rendering

CHAPTER 17

3. The two-view projection (Fig. 17.37) shows a ducting pipe. Construct a 3D model drawing of the pipe. Place in an **SW Isometric** view, add lighting to the scene and a material to the model and render.

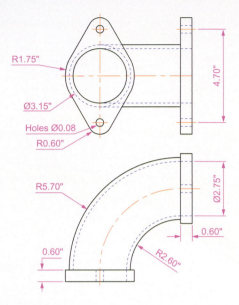

Fig. 17.37 Exercise 3 – details of shape and sizes

4. A point marking device is shown in two two-view projections (Fig. 17.38). The device is composed of three parts – a base, an arm and a pin. Construct a 3D model of the assembled device and add appropriate materials to each part. Then add lighting to the scene and render in an **SW Isometric** view (Fig. 17.39).

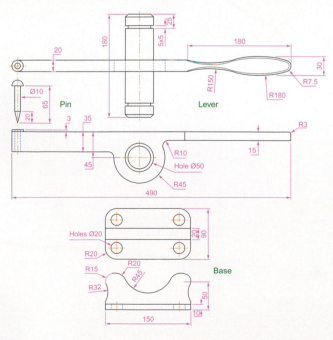

Fig. 17.38 Exercise 4 – details of shapes and sizes

Fig. 17.39 Exercise 4 – a rendering

5. A rendering of a 3D model drawing of the connecting device shown in the orthographic projection (Fig. 17.40) is given in Fig. 17.41. Construct the 3D model drawing of the device and add a suitable lighting to the scene.

Then place in the **ViewCube/Isometric** view, add a material to the model and render.

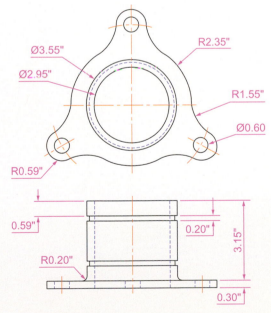

Ø3.55"
Ø2.95"
R2.35"
R1.55"
Ø0.60
R0.59"
0.59"
0.20"
3.15"
R0.20"
0.30"

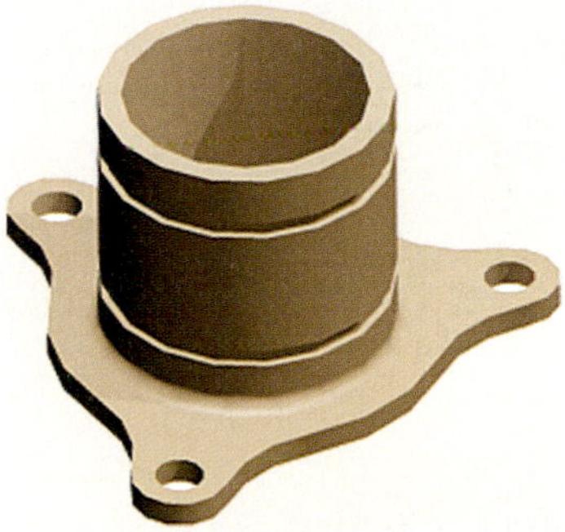

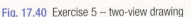

Fig. 17.41 Exercise 5 – a rendering

Fig. 17.40 Exercise 5 – two-view drawing

6. A fork connector and its rod are shown in a two-view projection (Fig. 17.42). Construct a 3D model drawing of the connector with its rod in position. Then add lighting to the scene, place in the **ViewCube/Isometric** viewing position, add materials to the model and render.

CHAPTER 17

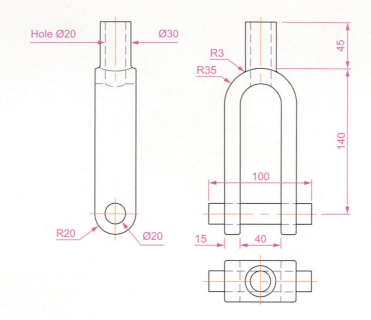

Fig. 17.42 Exercise 6

7. An orthographic projection of the parts of a lathe steady is given in Fig. 17.43. From the dimensions shown in the drawing, construct an assembled 3D model of the lathe steady.

When the 3D model has been completed, add suitable lighting and materials and render the model (Fig. 17.44).

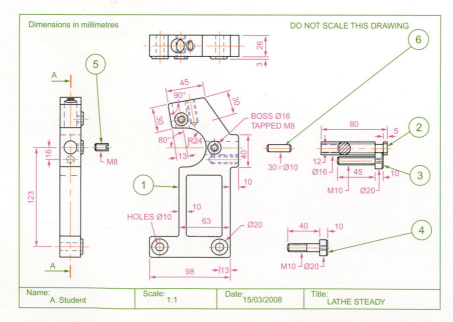

Fig. 17.43 Exercise 7 – details

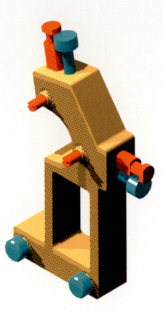

Fig. 17.44 Exercise 7 – a rendering

8. Construct suitable polylines to sizes of your own discretion in order to form the two surfaces to form the box shape shown in Fig. 17.45 with the aid of the **Rulesurf** tool. Add lighting and a material and render the surfaces so formed. Construct another three **Edgesurf** surfaces to form a lid for the box. Place the surface in a position above the box, add a material and render (Fig. 17.46).

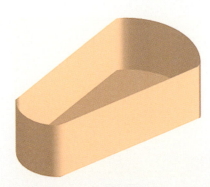

Fig. 17.45 Exercise 8 – the box

Fig. 17.46 Exercise 8 – the box and its lid

9. Fig. 17.47 shows a polyline for each of the 4 objects from which the surface shown in Fig. 17.48 was obtained. Construct the surface and shade in **Shades of Gray**.

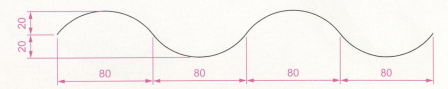

Fig. 17.47 Exercise 9 – one of the polylines from which the surface was obtained

CHAPTER 17

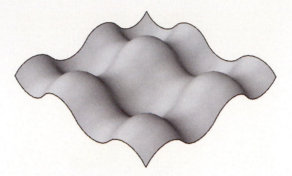

Fig. 17.48 Exercise 9

10. The surface model for this exercise was constructed from three **Edgesurf** surfaces working to the suggested objects for the surface as shown in Fig. 17.49. The sizes of the outlines of the objects in each case are left to your discretion. Fig. 17.50 shows the completed surface model. Fig. 17.51 shows the three surfaces of the model separated from each other.

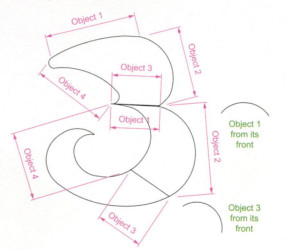

Fig. 17.49 Outlines for the three surfaces

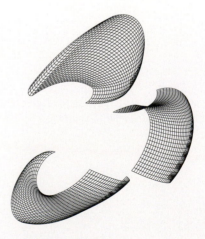

Fig. 17.50 Exercise 10

Fig. 17.51 The three surfaces

11. Fig. 17.52 shows in a **View Block/Isometric** view a semicircle of radius **25** constructed in the **View Cube/Top** view on a layer of colour **Magenta** with a semicircle of radius **75** constructed on the **View Block/Front** view with its left-hand end centred on the semicircle. Fig. 17.53 shows a surface constructed from the two semicircles in a **Visual Styles/Realistic** mode.

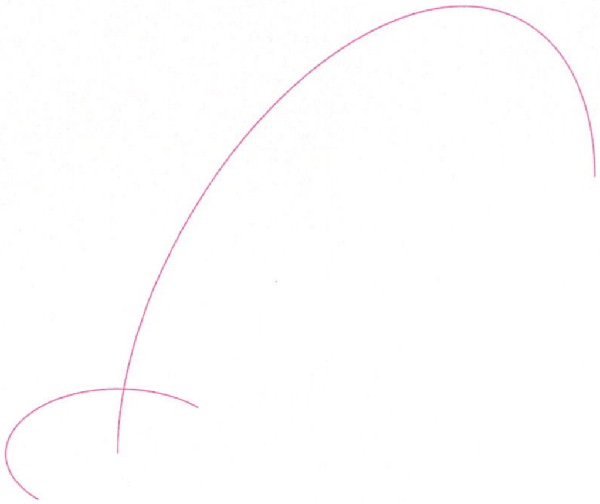

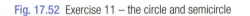

Fig. 17.52 Exercise 11 – the circle and semicircle

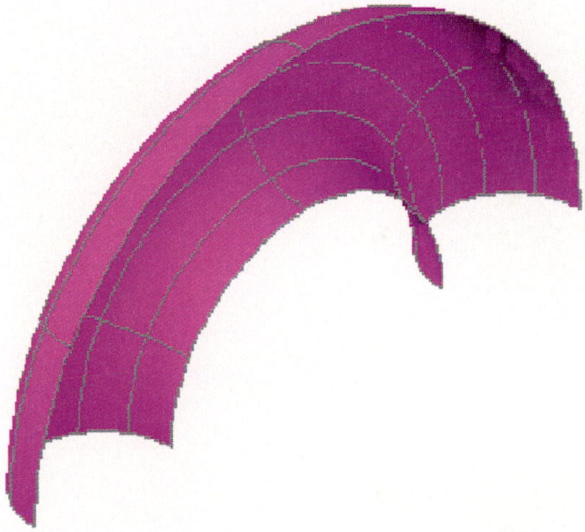

Fig. 17.53 Exercise 11

Editing 3D solid models

AIMS OF THIS CHAPTER

The aims of this chapter are:

1. To introduce the use of tools from the **Solid Editing** panel.
2. To show examples of a variety of 3D solid models.

The Solid Editing tools

The **Solid Editing** tools can be selected from the **Home/Solid Editing** panel (Fig. 18.1).

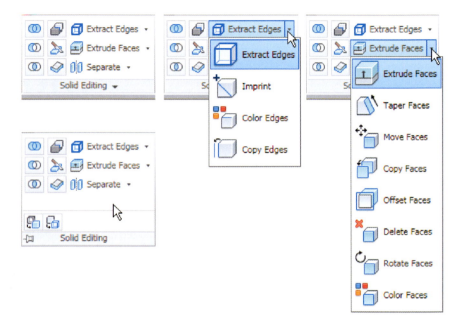

Fig. 18.1 The **Home/Solid Editing** panel

Examples of the results of using some of the **Solid Editing** tools are shown in this chapter. These tools are of value if the design of a 3D solid model requires to be changed (edited), although some have a value in constructing parts of 3D solids which cannot easily be constructed using other tools.

First example – Extrude faces tool (Fig. 18.3)

1. Set **ISOLINES to** 24.
2. In a **ViewCube/Right** view, construct a cylinder of radius **30** and height **30** (Fig. 18.3).
3. In a **ViewCube/Front** view, construct the pline (Fig. 18.2). Mirror the pline to the other end of the cylinder.

Fig. 18.3 First example – **Extrude faces tool**

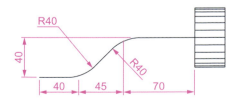

Fig. 18.2 First example – **Extrude faces tool** – first stages

4. In a **ViewCube/Top** view, move the pline to lie central to the cylinder.
5. Place the screen in a **ViewCube/Isometric view**.
6. *Click* the **Extrude faces** tool icon in the **Home/Solid Editing** panel (Fig. 18.1). The command line shows:

```
Command: _solidedit
Solids editing automatic checking: SOLIDCHECK=1
Enter a solids editing option [Face/Edge/Body/
  Undo/eXit] <eXit>: _face
Enter a face editing option
[Extrude/Move/Rotate/Offset/Taper/Delete/Copy/
  coLor/mAterial/Undo/eXit] <eXit>: _extrude
Select faces or [Undo/Remove]: pick the cylinder 2
  faces found.
Select faces or [Undo/Remove/ALL]: enter r right-
  click
Remove faces or [Undo/Add/ALL]: right-click
Specify height of extrusion or [Path]: enter
  p (Path) right-click
Select extrusion path: pick the left-hand path
  pline
Solid validation started.
Solid validation completed.
Enter a face editing option [Extrude/Move/Rotate/
  Offset/Taper/Delete/Copy/coLor/mAterial/Undo/
  eXit] <eXit>: right-click
Command:
```

7. Repeat the operation using the pline at the other end of the cylinder as a path.
8. Add lights and a material and render the 3D model (Fig. 18.3).

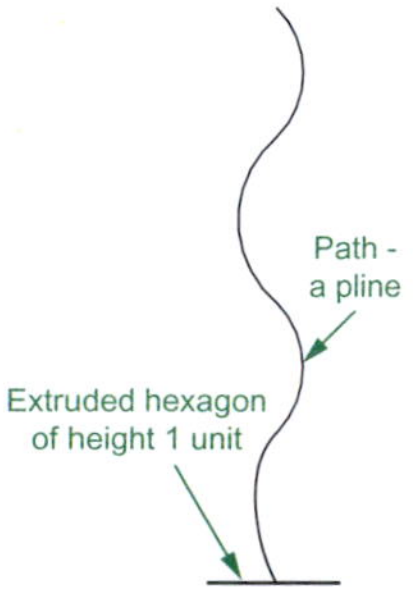

Path –
a pline

Extruded hexagon
of height 1 unit

Fig. 18.4 Second example – **Extrude faces tool** – pline for path

> **Note**
>
> Note the prompt line which includes the statement **SOLIDCHECK=1**. If the variable **SOLIDCHECK** is set on (to **1**) the prompt lines include the lines **SOLIDCHECK=1**, **Solid validation started** and **Solid validation completed**. If set to **0** these two lines do not show.

Second example – Extrude faces tool (Fig. 18.5)

1. Construct a hexagonal extrusion just **1** unit high in the **ViewCube/Top**.
2. Change to the **ViewCube/Front** and construct the curved pline (Fig. 18.4).

3. Back in the **Top** view, move the pline to lie central to the extrusion.
4. Place in the **ViewCube/Isometric** view and extrude the top face of the extrusion along the path of the curved pline.
5. Add lighting and a material to the model and render (Fig. 18.5).

> **Note**
>
> This example shows that a face of a 3D solid model can be extruded along any suitable path curve. If the polygon on which the extrusion had been based had been turned into a region, no extrusion could have taken place. The polygon had to be extruded to give a face to a 3D solid.

Third example – Move faces tool (Fig. 18.6)

1. Construct the 3D solid drawing shown in the left-hand drawing of Fig. 18.6 from three boxes which have been united using the **Union** tool.
2. *Click* on the **Move faces** tool in the **Home/Solid Editing** panel (see Fig. 18.1). The command line shows:

```
Command: _solidedit
[prompts] _face
Enter a face editing option
[prompts]: _move
Select faces or [Undo/Remove]: pick the model face
  4 face found.
Select faces or [Undo/Remove/ALL]: right-click
Specify a base point or displacement: pick
Specify a second point of displacement: pick
[further prompts]:
```

And the *picked* face is moved – right-hand drawing of Fig. 18.6.

Fig. 18.5 Second example – **Extrude faces tool**

Before
Move Faces

After
Move Faces

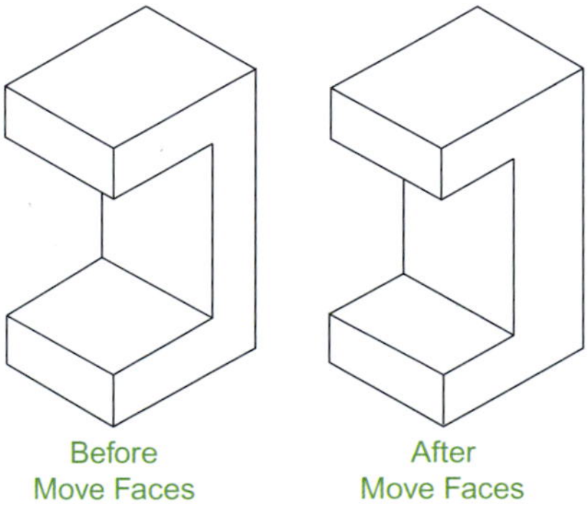

Fig. 18.6 Third example – **Solid, Move faces** tool

Fourth example – Offset faces (Fig. 18.7)

1. Construct the 3D solid drawing shown in the left-hand drawing of Fig. 18.7 from a hexagonal extrusion and a cylinder which have been united using the **Union** tool.

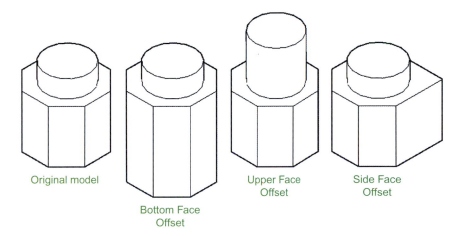

Original model

Bottom Face
Offset

Upper Face
Offset

Side Face
Offset

Fig. 18.7 Fourth example – **Offset faces tool**

2. *Click* on the **Offset faces** tool icon in the **Home/Solid Editing** panel (Fig. 18.1). The command line shows:

```
Command:_solidedit
[prompts]:_face
[prompts]
[prompts]:_offset
Select faces or [Undo/Remove]: pick the bottom
  face of the 3D model 2 faces found.
Select faces or [Undo/Remove/All]: enter r right-
  click
Select faces or [Undo/Remove/All]: pick
  highlighted faces other than the bottom face 2
  faces found, 1 removed
Select faces or [Undo/Remove/All]: right-click
Specify the offset distance: enter 30 right-click
```

3. Repeat, offsetting the upper face of the cylinder by **50** and the right-hand face of the lower extrusion by **15**.

The results are shown in Fig. 18.9.

Fifth example – Taper faces tool (Fig. 18.8)

1. Construct the 3D model as in the left-hand drawing of Fig. 18.8. Place in **ViewCube/Isometric** view.

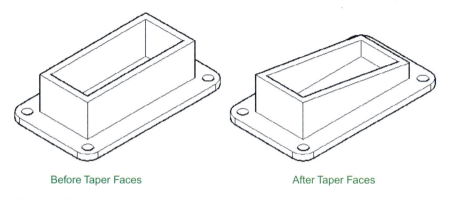

Before Taper Faces   After Taper Faces

Fig. 18.8 Fifth example – **Taper faces tool**

2. Call **Taper faces**. The command line shows:

```
Command:_solidedit
[prompts]:_face
[prompts]
[prompts]:_taper
Select faces or [Undo/Remove]: pick the upper face
  of the base 2 faces found.
Select faces or [Undo/Remove/All]: enter r right-
  click
Select faces or [Undo/Remove/All]: pick
  highlighted faces other than the upper face 2
  faces found, 1 removed
Select faces or [Undo/Remove/All]: right-click
Specify the base point: pick a point on left-hand
  edge of the face
Specify another point along the axis of tapering:
  pick a point on the right-hand edge of
  the face
Specify the taper angle: enter 10 right-click
```

And the selected face tapers as indicated in the right-hand drawing of Fig. 18.8.

Sixth example – Copy faces tool (Fig. 18.10)

1. Construct a 3D model to the sizes as given in Fig. 18.9.

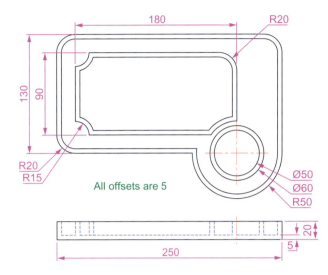

Fig. 18.9 Sixth example – **Copy Faces tool** – details of the 3D solid model

2. *Click* on the **Copy faces** tool in the **Home/Solid Editing** panel (Fig. 18.1). The command line shows:

```
Command:_solidedit
[prompts]:_face
[prompts]
[prompts]:_copy
Select faces or [Undo/Remove]: pick the upper face
  of the solid model 2 faces found.
Select faces or [Undo/Remove/All]: enter r right-
  click
Select faces or [Undo/Remove/All]: pick highlighted
  face not to be copied 2 faces found, 1 removed
Select faces or [Undo/Remove/All]: right-click
Specify a base point or displacement: pick
  anywhere on the highlighted face
Specify a second point of displacement: pick
  a point some 50 units above the face
```

3. Add lights and a material to the 3D model and its copied face and render (Fig. 18.10).

CHAPTER 18

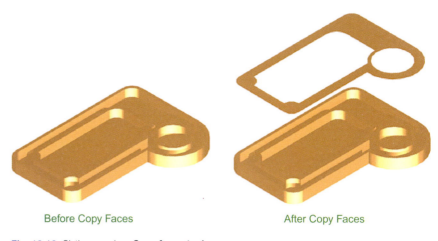

Before Copy Faces   After Copy Faces

Fig. 18.10 Sixth example – **Copy faces tool**

Seventh example – Color faces tool (Fig. 18.12)

1. Construct a 3D model of the wheel to the sizes as shown in Fig. 18.11.

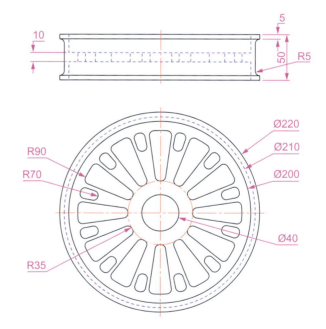

Fig. 18.11 Seventh example – **Color faces tool** – details of the 3D model

2. *Click* the **Color faces** tool icon in the **Home/Solid Editing** panel (Fig. 18.1). The command line shows:

```
Command:_solidedit
[prompts]:_face
```

```
[prompts]
[prompts]:_color
Select faces or [Undo/Remove]: pick the inner face
   of the wheel 2 faces found
Select faces or [Undo/Remove/All]: enter r right-
   click
Select faces or [Undo/Remove/All]: pick
   highlighted faces other than the required face 2
   faces found, 1 removed
Enter new color <ByLayer>: enter 1 (which is red)
   right-click
```

3. Add lights and a material to the edited 3D model and render (Fig. 18.12).

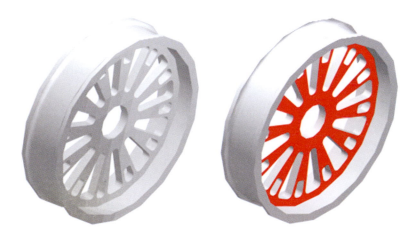

Fig. 18.12 Seventh example – **Color faces tool**

Examples of more 3D models

The following 3D models can be constructed in the **3d acadiso.dwt** screen. The descriptions of the stages needed to construct them have been reduced from those given in earlier pages, in the hope that readers have already acquired a reasonable skill in the construction of such drawings.

First example (Fig. 18.14)

1. **Front** view. Construct the three extrusions for the back panel and the two extruding panels to the details given in Fig. 18.13.

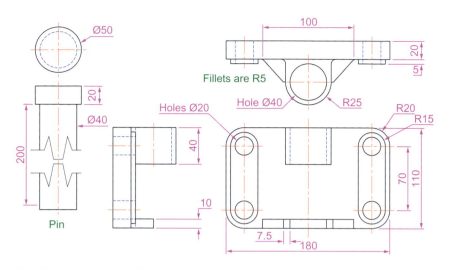

Fig. 18.13 First example – **3D models** – details of sizes and shapes

Fig. 18.14 First example – **3D models**

2. **Top** view. Move the two panels to the front of the body and union the three extrusions. Construct the extrusions for the projecting parts holding the pin.
3. **Front** view. Move the two extrusions into position and union them to the back.
4. **Top** view. Construct two cylinders for the pin and its head.
5. **Top** view. Move the head to the pin and union the two cylinders.
6. **Front** view. Move the pin into its position in the holder. Add lights and materials.
7. **Isometric** view. Render. Adjust lighting and materials as necessary (Fig. 18.14).

Second example (Fig. 18.16)

1. **Top**. (Fig. 18.15) Construct polyline outlines for the body extrusion and the solids of revolution for the two end parts. Extrude the body and subtract its hole and using the **Revolve** tool form the two end solids of revolution.
2. **Right**. Move the two solids of revolution into their correct positions relative to the body and union the three parts. Construct a cylinder for the hole through the model.
3. **Front**. Move the cylinder to its correct position and subtract from the model.
4. **Top**. Add lighting and a material.
5. **Isometric**. Render (Fig. 18.16).

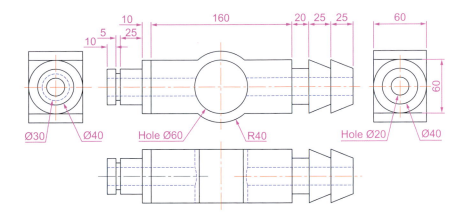

Fig. 18.15 Second example – **3D models** dimensions

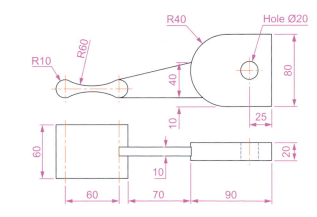

Fig. 18.16 Second example – **3D models**

Third example (Fig. 18.18)

1. **Front**. Construct the three plines needed for the extrusions of each part of the model (details Fig. 18.17). Extrude to the given heights. Subtract the hole from the **20** high extrusion.

Fig. 18.17 Third example – **3D models** – details of shapes and sizes

2. **Top**. Move the **60** extrusion and the **10** extrusion into their correct positions relative to the **20** extrusion. With **Union** form a single 3D model from the three extrusions.
3. Add suitable lighting and a material to the model.
4. **Isometric**. Render (Fig. 18.18).

Fig. 18.18 Third example – **3D models**

Fourth example (Fig. 18.19)

1. **Front**. Construct the polyline – left-hand drawing of Fig. 18.19.

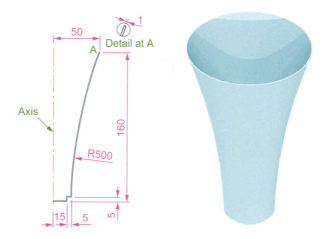

Fig. 18.19 Fourth example – **3D models**

2. With the **Revolve** tool from the **Home/3D Modeling** panel construct a solid of revolution from the pline.
3. **Top**. Add suitable lighting a coloured glass material.
4. **Isometric**. Render – right-hand illustration of Fig. 18.19.

Exercises

Methods of constructing answers to the following exercises can be found in the free website:

http://books.elsevier.com/companions/978-0-08-096575-8

1. Working to the shapes and dimensions as given in the orthographic projection of Fig. 18.20, construct the exploded 3D model as shown in Fig. 18.21. When the model has been constructed add suitable lighting and apply materials, followed by rendering.

2. Working to the dimensions given in the orthographic projections of the three parts of this 3D model (Fig. 18.22), construct the assembled as shown in the rendered 3D model (Fig. 18.23). Add suitable lighting and materials, place in one of the isometric viewing position and render the model.

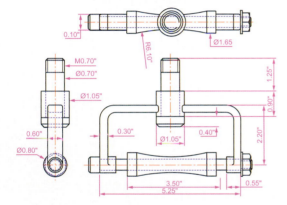
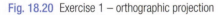

Fig. 18.20 Exercise 1 – orthographic projection

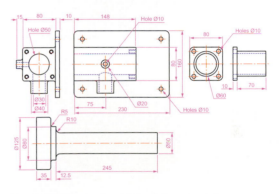
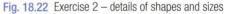

Fig. 18.22 Exercise 2 – details of shapes and sizes

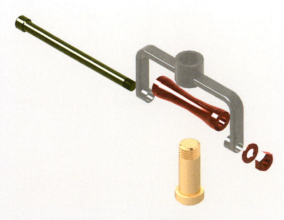

Fig. 18.21 Exercise 1 – rendered 3D model

Fig. 18.23 Exercise 2

CHAPTER 18

3. Construct the 3D model shown in the rendering (Fig. 18.24) from the details given in the parts drawing (Fig. 18.25).

Fig. 18.24 Exercise 3

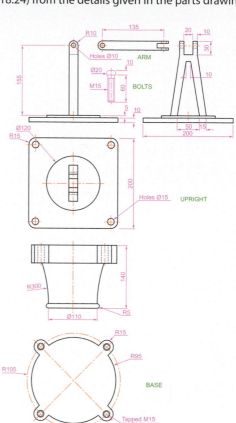

Fig. 18.25 Exercise 3 – the parts drawing

4. A more difficult exercise.

A rendered 3D model of the parts of an assembly is shown in Fig. 18.26.

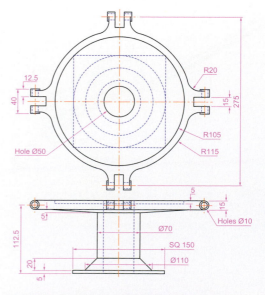

Fig. 18.26 Exercise 4 – first orthographic projection

Working to the details given in the three orthographic projections (Figs 18.26–18.28), construct the two parts of the 3D model, place them in suitable positions relative to each other, add lighting and materials and render the model (Fig. 18.29).

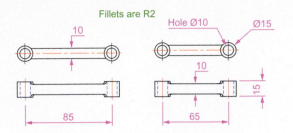

Fig. 18.27 Exercise 4 – third orthographic projections

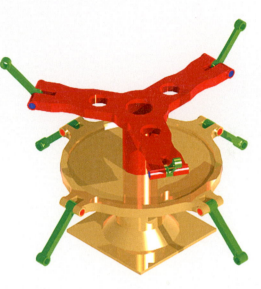

Fig. 18.29 Exercise 4

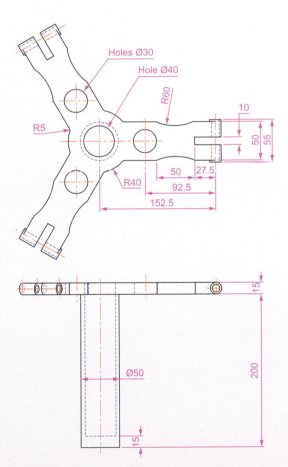

Fig. 18.28 Exercise 4 – second orthographic projection

Other features of 3D modeling

AIMS OF THIS CHAPTER

The aims of this chapter are:

1. To give a further example of placing raster images in an AutoCAD drawing.
2. To give examples of methods of printing or plotting not given in previous chapters.
3. To give examples of polygonal viewports.

Raster images in AutoCAD drawings

Example – Raster image in a drawing (Fig. 19.5)

This example shows the raster file **Fig05.bmp** of the 3D model constructed to the details given in Fig. 19.1.

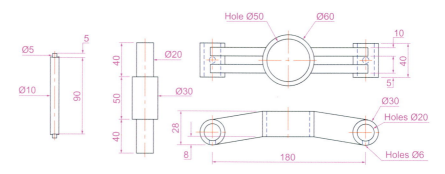

Fig. 19.1 **Raster image in a drawing** – drawings into which file is to be inserted

Raster images are graphics images in files with file names ending with the extensions ***.bmp**, ***.pcx**, ***.tif** and the like. The types of graphics files which can be inserted into AutoCAD drawings can be seen by first *clicking* on the **External References Palette** icon in the **View/Palettes** panel (Fig. 19.2).

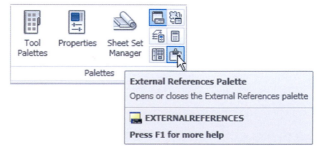

Fig. 19.2 Selecting **External Reference Palette** from the **View/Palettes** panel

Fig. 19.3 The External References palette

Then selecting **Attach Image…** from the popup menu brought down with a *click* on the left-hand icon at the top of the palette which brings the **Select Image File** dialog (Fig. 19.3) which brings the **Select Reference File** dialog on screen (Fig. 19.4).

In the dialog select the required raster file (in this example **Fig05.bmp**) and *click* the **Open** button. The **Attach Image** dialog appears showing

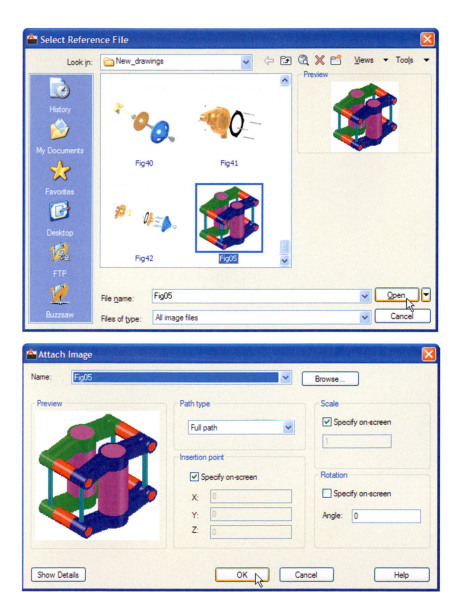

Fig. 19.4 **Raster image in a drawing** – the **Select Reference File** and **Attach Image** dialogs

the selected raster image. If satisfied *click* the **OK** button. The dialog disappears and the command line shows:

```
Command: _IMAGEATTACH
Specify insertion point <0,0>: pick
Base image size: Width: 1.000000, Height:
  1.041958, Millimeters
Specify scale factor <1>: enter 60 right-click
Command:
```

And the image is attached on screen at the *picked* position.

How to produce a raster image

1. Construct the 3D model to the shapes and sizes given in Fig. 19.1 working in four layers, each of a different colour.
2. Place in the **ViewCube/Isometric** view.
3. Shade the 3D model in **Realistic** visual style.
4. **Zoom** the shaded model to a suitable size and press the **Print Scr** key of the keyboard.
5. Open the Windows **Paint** application and *click* **Edit** in the menu bar, followed by another *click* on **Paste** in the drop-down menu. The whole AutoCAD screen which includes the shaded 3D assembled model appears.
6. *Click* the **Select** tool icon in the toolbar of **Paint** and window the 3D model. Then *click* **Copy** in the **Edit** drop-down menu.
7. *Click* **New** in the **File** drop-down menu, followed by a *click* on **No** in the warning window which appears.
8. *Click* **Paste** in the **Edit** drop-down menu. The shaded 3D model appears. *Click* **Save As…** from the **File** drop-down menu and save the bitmap to a suitable file name – in this example **Fig05.bmp**.
9. Open the orthographic projection drawing (Fig. 19.1) in AutoCAD.
10. Following the details given on page 386 attach **Fig05.bmp** to the drawing at a suitable position (Fig. 19.5).

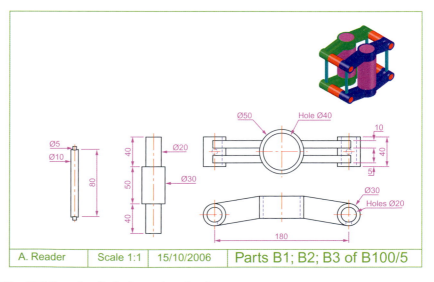

Fig. 19.5 Example – **Raster image in a drawing**

> **Notes**
>
> 1. It will normally be necessary to *enter* a scale in response to the prompt lines, otherwise the raster image may appear very small on screen. If it does it can be zoomed anyway.
>
> 2. Place the image in position in the drawing area. In Fig. 19.5 the orthographic projections have been placed within a margin and a title block has been added.

Hardcopy (prints or plots on paper) from a variety of different types of AutoCAD drawings of 3D models can be obtained. Some of this variety has already been shown in Chapter 15.

Printing/Plotting

First example – Printing/Plotting (Fig. 19.10)

If an attempt is made to print a multiple viewport screen in Model Space with all viewport drawings appearing in the plot, only the current viewport will be printed. To print or plot all viewports:

1. Open a four-viewport screen of the assembled 3D model shown in the first example (Fig. 19.5).
2. Make a new layer **vports** of colour **green**. Make this layer current.
3. *Click* the **MODEL** button in the status bar (Fig. 19.6). The **Page Setup Manager** dialog appears (Fig. 19.7). *Click* its **Modify…** button and the **Page Setup – Layout1** dialog appears (Fig. 19.8).

Fig. 19.6   First example – the **MODEL** button in the status bar

4. Make settings as shown and *click* the dialog's **OK** button, the **Page Setup Manager** dialog reappears showing the new settings. *Click* its **Close** button. The current viewport appears.
5. Erase the green outline and the viewport is erased.
6. At the command line:

```
Command: enter mv
MVIEW
```

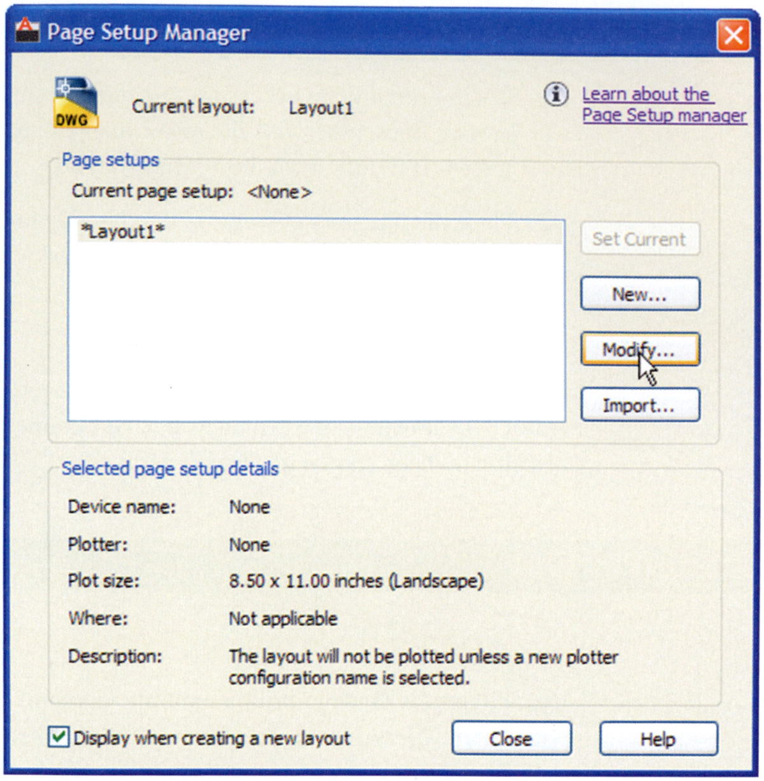

Fig. 19.7 The **Page Setup Manager** dialog

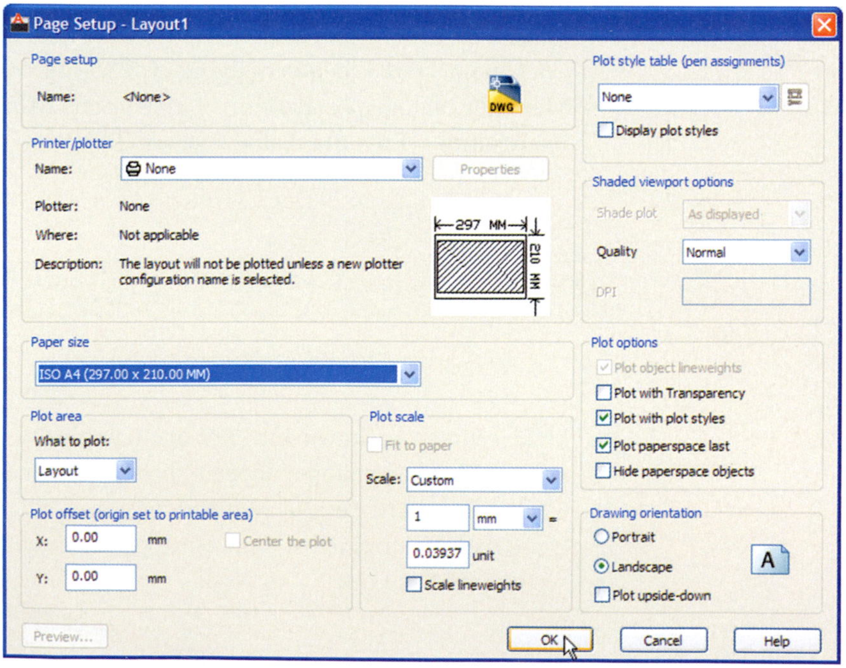

Fig. 19.8 The **Page Setup – Layout1** dialog

```
Specify corner of viewport or
[ON/OFF/Fit/Shadeplot/Lock/Object/Polygonal/
  Restore/LAyer/2/3/4] <Fit>: enter r right-click
Enter viewport configuration name or [?]
  <*Active>: right-click
Specify first corner or [Fit] <Fit>: right-
  click
Command:
```

7. Turn layer **vports** off.
8. *Click* the **PAPER** button (note it changes from **MODEL**) and the current viewport changes to a model view. In each viewport in turn change the settings from the **3D Navigation** drop-down to **Front**, **Top**, **Right** and **SW isometric**. *Click* the **MODEL** button. It changes to **PAPER** and the screen reverts to **Pspace**.
9. *Click* the **Plot** tool icon in the **Quick Access** bar (Fig. 19.9). A **Plot** dialog appears.

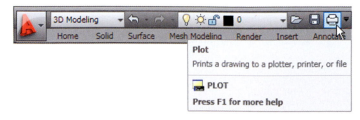

Fig. 19.9 The **Plot** tool icon in the **Quick Access** toolbar

10. Check in the dialog that the settings for the printer/plotter is correct and the paper size is also correct.
11. *Click* the **Preview** button. The full preview of the plot appears (Fig. 19.10).
12. *Right-click* anywhere in the drawing and *click* on **Plot** in the right-click menu which then appears.
13. The drawing plots (or prints).

Second example – Printing/Plotting (Fig. 19.11)

1. Open the orthographic drawing with its raster image (Fig. 19.5).
2. While still in **Model Space** *click* the **Plot** tool icon. The **Plot** dialog appears. Check that the required printer/plotter and paper size have been chosen.
3. *Click* the **Preview** button.

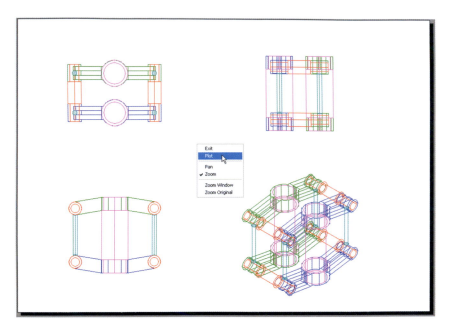

Fig. 19.10 First example – **Printing/Plotting**

4. If satisfied with the preview (Fig. 19.11), *right-click* and in the menu which appears *click* the name **Plot**. The drawing plots.

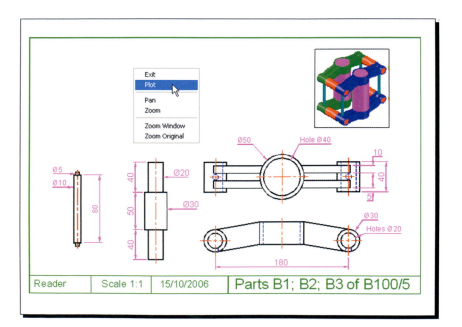

Fig. 19.11 Second example – **Printing/Plotting**

Third example – Printing/Plotting (Fig. 19.12)

1. Open the 3D model drawing of the assembly shown in Fig. 19.10 in a single **ViewCube/Isometric** view.
2. While in **MSpace**, *click* the **Plot** tool icon. The **Plot** dialog appears.
3. Check that the plotter device and sheet sizes are correct. *Click* the **Preview** button.
4. If satisfied with the preview (Fig. 19.12), *right-click* and *click* on **Plot** in the menu which appears. The drawing plots.

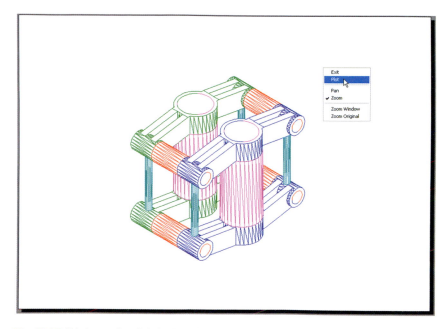

Fig. 19.12   Third example – Printing/Plotting

Fourth example – Printing/Plotting (Fig. 19.13)

Fig. 19.13 shows a **Plot Preview** of the 3D solid model (Fig. 18.29).

Polygonal viewports (Fig. 19.12)

The example to illustrate the construction of polygonal viewports is based upon Exercise 6. When the 3D model for this exercise has been completed in **Model Space**:

1. Make a new layer **vports** of colour **blue** and make it current.
2. Using the same methods as described for the first example of printing/ plotting produce a four-viewport screen of the model in **Pspace**.

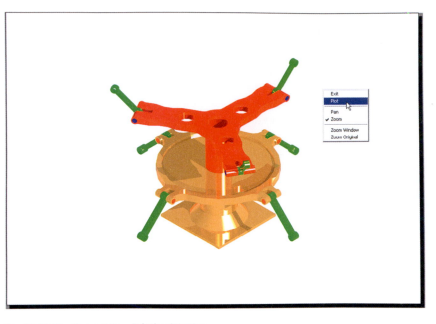

Fig. 19.13 Fourth example – **Printing/Plotting**

3. Erase the viewport with a *click* on its bounding line. The outline and its contents are erased.
4. *Click* the **Model** button. With a *click* in each viewport in turn and using the **ViewCube** settings set viewports in **Front**, **Right**, **Top** and **Isometric** views, respectively.
5. **Zoom** each viewport to **All**.
6. *Click* the **Layout1** button to turn back to **PSpace**.
7. Enter **mv** at the command line, which shows:

```
Command: enter mv right-click
MVIEW
[prompts]: enter p (Polygonal) right-click
Specify start point: In the top right viewport
  pick one corner of a square
Specify next point or [Arc/Close/Length/Undo]:
  pick next corner for the square
Specify next point or [Arc/Close/Length/Undo]:
  pick next corner for the square
Specify next point or [Arc/Close/Length/Undo]:
  enter c (Close)right-click
Regenerating model.
Command:
```

And a square viewport outline appears in the top right viewport within which is a copy of the model.

8. Repeat in each of the viewports with different shapes of polygonal viewport outlines (Fig. 19.14).

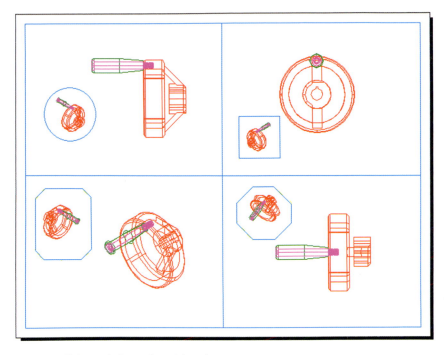

Fig. 19.14 **Polygonal viewports** – plot preview

9. *Click* the **Model** button.
10. In each of the polygonal viewports make a different isometric view. In the bottom right polygonal viewport change the view using the **3D Orbit** tool.
11. Turn the layer **vports** off. The viewport borders disappear.
12. *Click* the **Plot** icon. Make plot settings in the **Plot** dialog. *Click on* the **Preview** button of the **Plot** dialog. The **Preview** appears (Fig. 19.15).

The Navigation Wheel

The **Navigation Wheel** can be called from the **View/Navigate** panel as shown in Fig. 19.14. The reader is advised to experiment with the **Navigation Wheel** (Fig. 19.16).

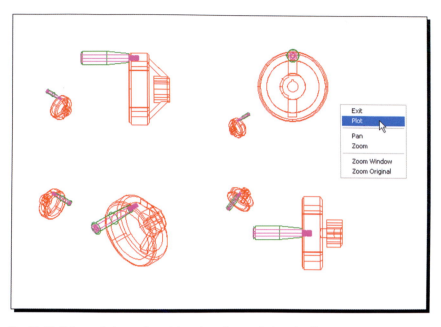

Fig. 19.15 **Polygonal viewports** – plot preview after **vports** layer is off

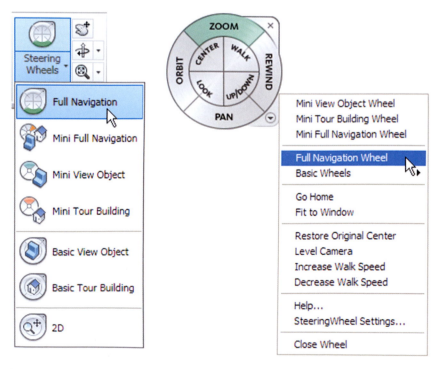

Fig. 19.16 The Navigation Wheel

The Mesh tools

Fig. 19.17 shows a series of illustrations showing the actions of the **Mesh** tools and the three 3D tools – **3dmove**, **3dscale** and **3drotate**. The illustrations show:

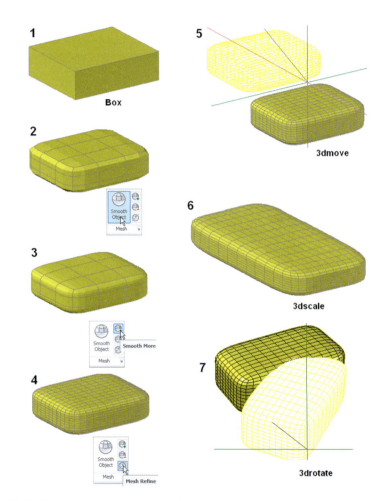

Fig. 19.17 Mesh: 3dmove, 3dscale and 3drotate tools

1. A box constructed using the **Box** tool.
2. The box acted upon by the **Smooth Object** tool from the **Home/Mesh** panel.
3. The box acted upon by the **Smooth Mesh** tool.
4. The box acted upon by the **Mesh Refine** tool.
5. The **Smooth refined** box acted upon by the **3dmove** tool.
6. The **Smooth Refined** box acted upon by the **3dscale** tool.
7. The **Smooth Refined** box acted upon by the **3drotate** tool.

Exercises

Methods of constructing answers to the following exercises can be found in the free website:

http://books.elsevier.com/companions/978-0-08-096575-8.

1. Working to the shapes and sizes given in Fig. 19.18, construct an assembled 3D model drawing of the spindle in its two holders, add lighting and apply suitable materials and render (Fig. 19.19).

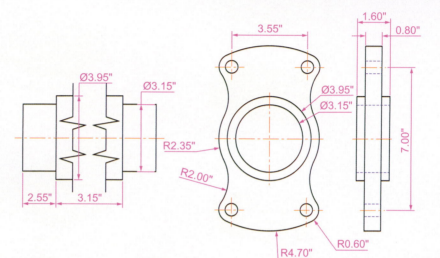

Fig. 19.18 Exercise 1 – details of shapes and sizes

model. Construct the 3D model, add lighting, apply suitable materials and render.

Fig. 19.19 Exercise 1

2. Fig. 19.20 shows a rendering of the model for this exercise and Fig. 19.21, an orthographic projection, giving shapes and sizes for the

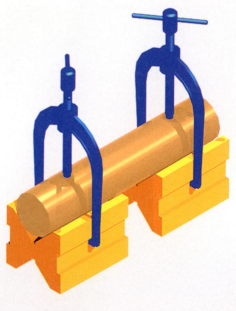

Fig. 19.20 Exercise 2

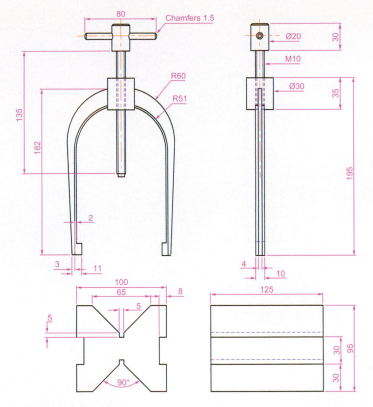

Fig. 19.21 Exercise 2 – orthographic projection

3. Construct a 3D model drawing to the details given in Fig. 19.22. Add suitable lighting and apply a material, then render as shown in Fig. 19.23.

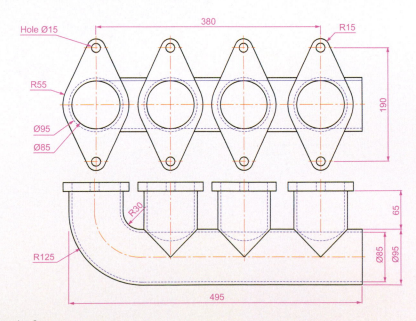

Fig. 19.22 Exercise 3

Fig. 19.23 Exercise 3 – **ViewCube/Isometric** view

4. Construct an assembled 3D model drawing working to the details given in Fig. 19.24. When the 3D model drawing has been constructed disassemble the parts as shown in the given exploded isometric drawing (Fig. 19.25).

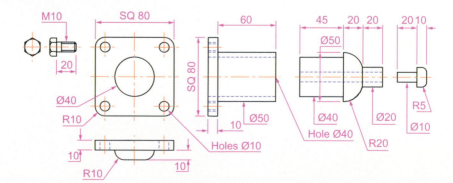

Fig. 19.24 Exercise 4 – details of shapes and sizes

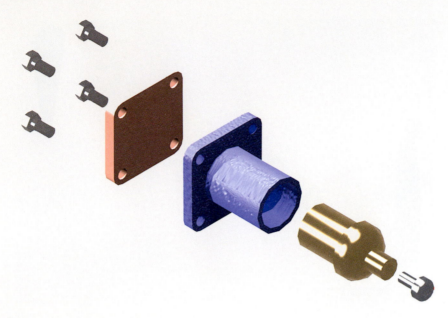

Fig. 19.25 Exercise 5 – an exploded rendered model

5. Working to the details shown in Fig. 19.26, construct an assembled 3D model, with the parts in their correct positions relative to each other. Then separate the parts as shown in the 3D rendered model drawing (Fig. 19.27). When the 3D model is complete add suitable lighting and materials and render the result.

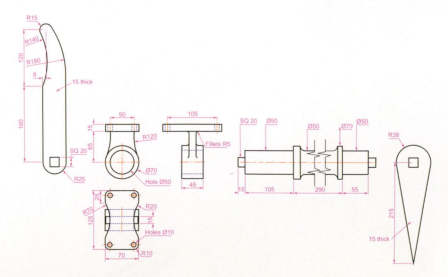

Fig. 19.26 Exercise 5 – details drawing

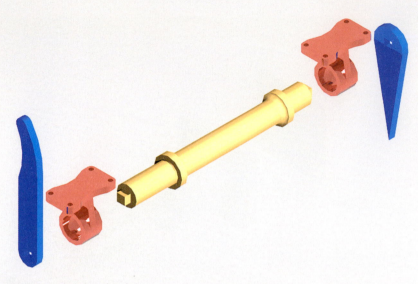

Fig. 19.27 Exercise 5 – exploded rendered view

6.   Working to the details shown in Fig. 19.28, construct a 3D model of the parts of the wheel with its handle. Two renderings of 3D models of the rotating handle are shown in Fig. 19.29, one with its parts assembled, the other with the parts in an exploded position relative to each other.

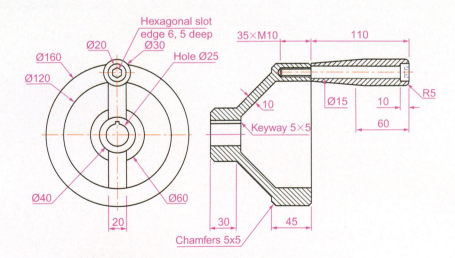

Fig. 19.28 Exercise 6 – details drawing

Fig. 19.29 Exercise 6 – renderings

Part 3

Internet tools and Design

Internet tools and Help

AIM OF THIS CHAPTER

The aim of this chapter is to introduce the tools which are available in AutoCAD 2011, which make use of facilities available on the World Wide Web (WWW).

Emailing drawings

As with any other files which are composed of data, AutoCAD drawings can be sent by email as attachments. If a problem of security of the drawings is involved they can be encapsulated with a password as the drawings are saved prior to being attached in an email. To encrypt a drawing with a password, *click* **Tools** in the **Save Drawing As** dialog and from the popup list which appears *click* **Security Options…** (Fig. 20.1).

Fig. 20.1 Selecting **Security Options** in the **Save Drawing As** dialog

Then in the **Security Options** dialog which appears (Fig. 20.2), *enter* a password in the **Password or phrase to open this drawing** field, followed by a *click* on the **OK** button. After *entering* a password *click* the **OK** button and *enter* the password in the **Confirm Password** dialog which appears.

Fig. 20.2 *Entering* and confirming a password in the **Security Options** dialog

The drawing then cannot be opened until the password is *entered* in the **Password** dialog which appears when an attempt is made to open the drawing by the person receiving the email (Fig. 20.3).

Password   ✕

Enter password to open drawing:
C:\Acad2009\Chapter19\inserts\Fig01.dwg

•••••••••••|

OK   Cancel

Fig. 20.3 The **Password** dialog appearing when a password encrypted drawing is about to be opened

There are many reasons why drawings may require to be password encapsulated in order to protect confidentiality of the contents of drawings.

Creating a web page (Fig. 20.5)

To create a web page which includes AutoCAD drawings first *left-click* **Publish to Web...** in the **File** drop-down menu (Fig. 20.4).

A series of **Publish to Web** dialogs appear, some of which are shown here in Figs 20.5–20.7. After making entries in the dialogs which come on screen after each **Next** button is *clicked*, the resulting web page such as that shown in Fig. 20.7 will be seen. A *double-click* in any of the thumbnail views in this web page and another page appears showing the selected drawing in full.

Fig. 20.4 The **Publish to Web** tool in the **File** drop-down menu

Publish to Web - Create Web Page   ✕

- Begin
- ▶ Create Web Page
- Edit Web Page
- Describe Web Page
- Select Image Type
- Select Template
- Apply Theme
- Enable i-drop
- Select Drawings
- Generate Images
- Preview and Post

Your Web page and its configuration file are stored in a directory in your file system to enable future editing and posting. The name of this directory is the same as the name of your Web page. You can choose the location (parent directory) where this folder is created.

Specify the name of your Web page (do not include a file extension).

62 Pheasant Drive

Specify the parent directory in your file system where the Web page folder will be created.

C:\Acad2010\Book ...\62 Pheasant Drive   [...]

Provide a description to appear on your Web page.

A series of drawings connected with the building of an extension to the building already at 62 Pheasant Drive

< Back   Next >   Cancel

Fig. 20.5 The **Publish to Web – Create Web Page** dialog

CHAPTER 20

Fig. 20.6 The Publish to Web – Select Template dialog

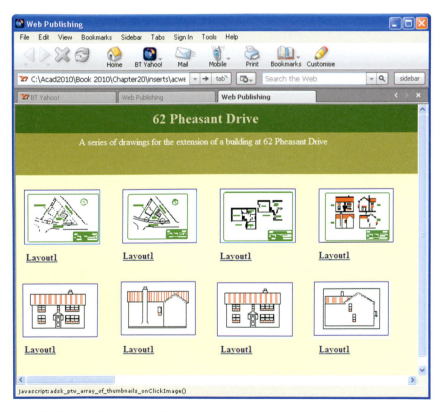

Fig. 20.7 The **Web Publishing – Windows Internet Explorer** page

The eTransmit tool

At the command line *enter* **eTransmit**. The **Create Transmittal** dialog appears (Fig. 20.8). The transmittal shown in Fig. 20.8 is the drawing on screen at the time the transmittal was made plus a second drawing. Fill in details as necessary. The transmittal is saved as a standard **zip** file.

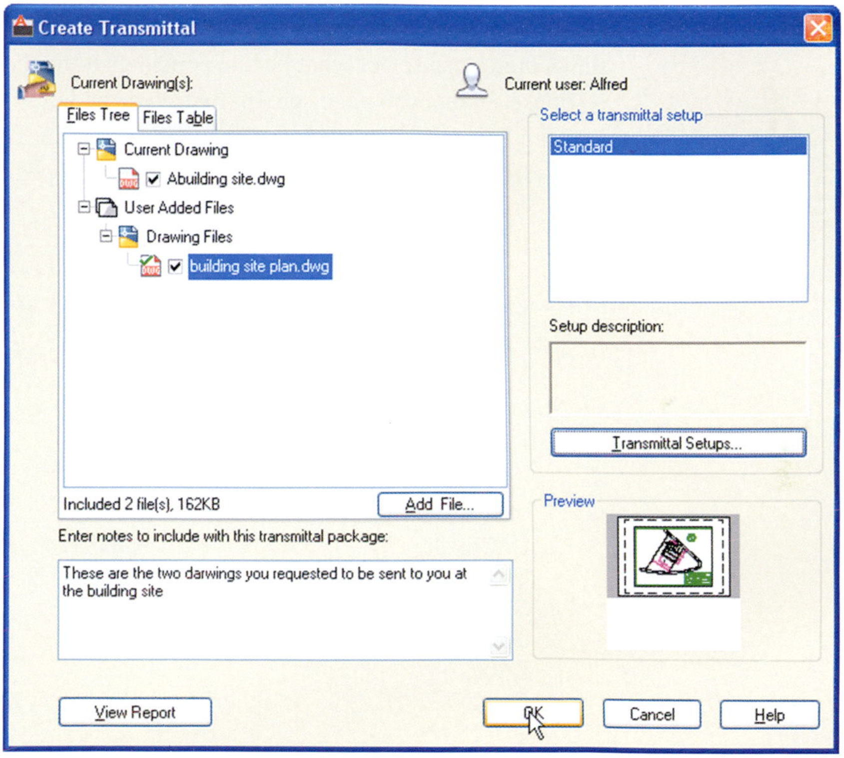

Fig. 20.8 The **Create Transmittal** dialog

> **Note**
>
> There is no icon for **eTransmit** in the ribbon panels.

Help

Fig. 20.9 shows a method of getting help. In this example help on using the **Break** tool is required. *Enter* **Break** in the **Search** field, followed

Fig. 20.9 Help for **Break**

by a *click* on the **Click here to access the help** button (Fig. 20.9). The **AutoCAD 2011 Help** web page appears (Fig. 20.10) appears from which the operator can select what he/she considers to be the most appropriate response. In the web page screen, first *click* the letter **B** in the **Command** list (Fig. 20.10). A list of commands with the initial **B** appears (Fig. 20.11). C*lick* **BREAK** in this list. The **Help** for **Break** appears (Fig. 20.12).

Commands

3D A B C D E F G H I J K L M N O P Q R S T
U V W X Y Z

Fig. 20.10 *Click* a **Commands** letter in the **AutoCAD 2011 Help** web page

B commands

BACTION	BCPARAMETER	BOX
BACTIONBAR	BCYCLEORDER	BPARAMETER
BACTIONSET	BEDIT	BREAK
BACTIONTOOL	BESETTINGS	BREP
BASE	BGRIPSET	BROWSER
BASSOCIATE	BHATCH	BSAVE
BATTMAN	BLIPMODE	BSAVEAS
BATTORDER	BLOCK	BTABLE
BAUTHORPALETTE	BLOCKICON	BTESTBLOCK
BAUTHORPALETTECLOSE	BLOOKUPTABLE	BVHIDE
BCLOSE	BMPOUT	BVSHOW
BCONSTRUCTION	BOUNDARY	BVSTATE

Fig. 20.11 *Click* the command name in the list which appears

The New Features Workshop

Click the down pointing arrow to the right of the **?** icon and select **New Features Workshop** from the menu which appears (Fig. 20.13) The **New Features Workshop** web page appears (Fig. 20.14) from which a selection of new features can be selected.

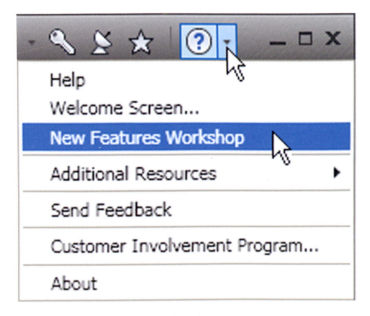

Fig. 20.12 The **AutoCAD 2011 AutoCAD Help** web page showing help for **Break**

Fig. 20.13 Select **New Features Workshop** from the arrow

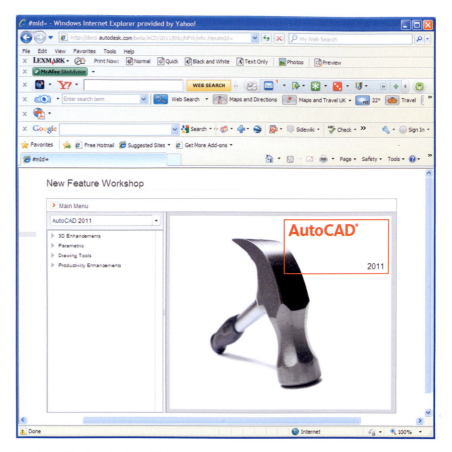

Fig. 20.14 The **New Features Workshop** web page

Design and AutoCAD 2011

AIMS OF THIS CHAPTER

The aims of this chapter are:

1. To describe reasons for using AutoCAD.
2. To describe methods of designing artefacts and the place of AutoCAD in the design process.
3. To list the system requirements for running AutoCAD 2011 software.
4. To list some of the enhancements in AutoCAD 2011.

Ten reasons for using AutoCAD

1. A CAD software package such as AutoCAD 2011 can be used to produce any form of technical drawing.
2. Technical drawings can be produced much more speedily using AutoCAD than when working manually – probably as much as 10 times as quickly when used by skilled AutoCAD operators.
3. Drawing with AutoCAD is less tedious than drawing by hand – features such as hatching, lettering and adding notes are easier, quicker and indeed more accurate to construct.
4. Drawings or parts of drawings can be moved, copied, scaled, rotated, mirrored and inserted into other drawings without having to redraw.
5. AutoCAD drawings can be saved to a file system without necessarily having to print the drawing. This can save the need for large paper drawing storage areas.
6. The same drawing or part of a drawing need never be drawn twice, because it can be copied or inserted into other drawings with ease. A basic rule when working with AutoCAD is *Never draw the same feature twice*.
7. New details can be added to drawings or be changed within drawings without having to mechanical erase the old detail.
8. Dimensions can be added to drawings with accuracy reducing the possibility of making errors.
9. Drawings can be plotted or printed to any scale without having to redraw.
10. Drawings can be exchanged between computers and/or emailed around the world without having to physically send the drawing.

The place of AutoCAD 2011 in designing

The contents of this book are only designed to help those who have a limited (or no) knowledge and skills of the construction of technical drawings using AutoCAD 2011. However it needs to be recognised that the impact of modern computing on the methods of designing in industry has been immense. Such features such as analysis of stresses, shear forces, bending forces and the like can be carried out more quickly and accurately using computing methods. The storage of data connected with a design and the ability to recover the data speedily are carried out much more easily using computing methods than prior to the introduction of computing.

AutoCAD 2011 can play an important part in the design process, because technical drawings of all types are necessary for achieving well designed artefacts whether it be an engineering component, a machine, a building, an electronics circuit or any other design project.

In particular, 2D drawings which can be constructed in AutoCAD 2011 are still of great value in modern industry. AutoCAD 2011 can also be used to produce excellent and accurate 3D models, which can be rendered to produce photographic like images of a suggested design. Although not dealt with in this book, data from 3D models constructed in AutoCAD 2011 can be taken for use in computer aided machining (CAM).

At all stages in the design process, either (or both) 2D or 3D drawings play an important part in aiding those engaged in designing to assist in assessing the results of their work at various stages. It is in the design process that drawings constructed in AutoCAD 2011 play an important part.

In the simplified design process chart shown in Fig. 21.1 an asterisk (*) has been shown against those features where the use of AutoCAD 2011 can be regarded as being of value.

A design chart (Fig. 21.1)

The simplified design chart Fig. 21.1 shows the following features:

Design brief: A design brief is a necessary feature of the design process. It can be in the form of a statement, but it is usually much more. A design

Fig. 21.1 A simplified design chart

brief can be a written report which not only includes a statement made of the problem which the design is assumed to be solving, but includes preliminary notes and drawings describing difficulties which may be encountered in solving the design and may include charts, drawings, costings, etc. to emphasise some of the needs in solving the problem for which the design is being made.

Research: The need to research the various problems which may arise when designing is often much more demanding than the chart (Fig. 21.1) shows. For example the materials being used may require extensive research as to costing, stress analysis, electrical conductivity, difficulties in machining or in constructional techniques and other such features.

Ideas for solving the brief: This is where technical, other drawings and sketches play an important part in designing. It is only after research that designers can ensure the brief will be fulfilled.

Models: These may be constructed models in materials representing the actual materials which have been chosen for the design, but in addition 3D solid model drawings, such as those which can be constructed in AutoCAD 2011, can be of value. Some models may also be made in the materials from which the final design is to be made so as to allow testing of the materials in the design situation.

Chosen solution: This is where the use of drawings constructed in AutoCAD 2011 is of great value. 2D and 3D drawings come into their own here. It is from such drawings that the final design will be manufactured.

Realisation: The design is made. There may be a need to manufacture a number of the designs in order to enable evaluation of the design to be fully assessed.

Evaluation: The manufactured design is tested in situations such as it is liable to be placed in use. Evaluation will include reports and notes which could include drawings with suggestions for amendments to the working drawings from which the design was realised.

Enhancements in AutoCAD 2011

AutoCAD 2011 contains many enhancements over previous releases, whether working in a **2D** or a **3D** workspace. Please note that not all the enhancements in AutoCAD 2011 are described in this introductory book. Among the more important enhancements are the following:

1. When first loaded, an **Initial Setup** dialog offers a **Welcome Screen** from which the operator can select from a variety of videos illustrating how different methods of drawing in both 2D and 3D can be used in AutoCAD 2011.

2. The **Ribbon** has been amended and brought up to date. A new ribbon **Hatch Creation** from which hatch tools can be chosen when hatching.

3. A new feature – the **Navigation Bar** has been introduced situated at the right-hand edge of the AutoCAD 2011 window. The tools in this bar are frequently used and can be assessed speedily from the navigation bar.

4. The **ViewCube** is now available in the **2D Drafting and Annotation** workspace.

5. A new workspace **3D Basic** has been introduced with its own ribbon showing basic 3D tools in its panels.

6. Any part of a drawing can be made partly transparent using the new tool **Transparency** in the **2D Drafting and Annotation** ribbon from the **Properties** panel.

7. The buttons in the status bar now include **Selection Cycling**, **Show/ Hide Transparency**, **3D Object Snap**, **Infer Restraints, Isolate Objects** and **Hardware Acceleration**. Some buttons in previous releases are no longer included in the status bar.

8. Two new commands – **Chamferedge** and **Filletedge** – allow modifications to chamfers and fillets.

9. 3D materials enhancements. New **Materials Browser** and **Materials Editor** palettes. Materials can be selected for assigning to 3D objects or can be *dragged* on the objects for assigning.

10. A larger number of materials available from several different folders.

11. Materials can be selected from other Autodesk software such as **Maya** or **3D Studio Max**.

12. 3D ribbon reorganised in the **3D Modeling** workspace.

System requirements for running AutoCAD 2011

Note: There are two editions of AutoCAD 2011 – 32 bit and 64 bit editions.

Operating system: Windows XP Professional, Windows XP Professional (×64 Edition), Windows XP Home Edition, Windows 2000 or Windows Vista 32 bit, Windows Vista 64 bit, Windows 7.

Microsoft Internet Explorer 7.0.

Processor: Pentium III 800 MHz or equivalent.

Ram: At least 128 MB.

Monitor screen: 1024 × 768 **VGA** with **True Colour** as a minimum.

Hard disk: A minimum of 300 MB.

Graphics card: An AutoCAD certified graphics card. Details can be found on the web page **AutoCAD Certified Hardware XML Database**.

Appendices

Appendix A
List of tools

Introduction

AutoCAD 2011 allows the use of over 1000 commands (or tools). A selection of the most commonly used from these commands (tools) is described in this appendix. Some of the commands described here have not been used in this book, because this book is an introductory text designed to initiate readers into the basic methods of using AutoCAD 2011. It is hoped the list will encourage readers to experiment with those tools not described in the book. The abbreviations for tools which have them are included in brackets after the tool name. Tool names can be *entered* at the command line in upper or lower case.

A list of 2D commands is followed by a listing of 3D commands. Internet commands are described at the end of this listing. It must be remembered that not all of the tools available in AutoCAD 2011 are shown here.

The abbreviations for the commands can be found in the file **acad.pgp** from the folder: **C:\Autodesk\AutoCAD_2011_English_Win_32bit_SLD\x86\acad\en-us\Acad\Program Files\Root\UserDataCache\Support**.

Not all of the commands have abbreviations.

2D commands

About – Brings the **About AutoCAD** bitmap on screen
Appload – Brings the **Load/Unload Applications** dialog to screen
Adcenter (dc) – Brings the **DesignCenter** palette on screen
Align (al) – Aligns objects between chosen points
Arc (a) – Creates an arc
Area – States in square units of the area selected from a number of points
Array (ar) – Creates **Rectangular** or **Polar** arrays in 2D
Ase – Brings the **dbConnect Manager** on screen
Attdef – Brings the **Attribute Definition** dialog on screen
Attedit – Allows editing of attributes from the Command line
Audit – Checks and fixes any errors in a drawing
Autopublish – Creates a **DWF** file for the drawing on screen
Bhatch (h) – Brings the **Boundary Hatch** dialog on screen
Block – Brings the **Block Definition** dialog on screen
Bmake (b) – Brings the **Block Definition** dialog on screen
Bmpout – Brings the **Create Raster File** dialog
Boundary (bo) – Brings the **Boundary Creation** dialog on screen
Break (br) – Breaks an object into parts

Cal – Calculates mathematical expressions

Chamfer (cha) – Creates a chamfer between two entities

Chprop (ch) – Brings the **Properties** window on screen

Circle (c) – Creates a circle

Copytolayer – Copies objects from one layer to another

Copy (co) – Creates a single or multiple copies of selected entities

Copyclip (Ctrl+C) – Copies a drawing or part of a drawing for inserting into a document from another application

Copylink – Forms a link between an AutoCAD drawing and its appearance in another application such as a word processing package

Customize – Brings the **Customize** dialog to screen, allowing the customisation of toolbars, palettes, etc.

Dashboard – Has the same action as **Ribbon**

Dashboardclose – Closes the **Ribbon**

Ddattdef (at) – Brings the **Attribute Definition** dialog to screen

Ddatte (ate) – Edits individual attribute values

Ddcolor (col) – Brings the **Select Color** dialog on screen

Ddedit (ed) – The **Text Formatting** dialog box appears on selecting text

Ddim (d) – Brings the **Dimension Style Manager** dialog box on screen

Ddinsert (i) – Brings the **Insert** dialog on screen

Ddmodify – Brings the **Properties** window on screen

Ddosnap (os) – Brings the **Drafting Settings** dialog on screen

Ddptype – Brings the **Point Style** dialog on screen

Ddrmodes (rm) – Brings the **Drafting Settings** dialog on screen

Ddunits (un) – Brings the **Drawing Units** dialogue on screen

Ddview (v) – Brings the **View Manager** on screen

Del – Allows a file (or any file) to be deleted

Dgnexport – Creates a **MicroStation V8 dgn** file from the drawing on screen

Dgnimport – Allows a **MicroStation V8 dgn** file to be imported as an AutoCAD dwg file

Dim – Starts a session of dimensioning

Dimension tools – The **Dimension** toolbar contains the following tools – **Linear**, **Aligned**, **Arc Length**, **Ordinate**, **Radius**, **Jogged**, **Diameter**, **Angular**, **Quick Dimension**, **Baseline**, **Continue**, **Quick Leader**, **Tolerance**, **Center Mark**, **Dimension Edit**, **Dimension Edit Text**, **Update** and **Dimension Style**

Dim1 – Allows the addition of a single addition of a dimension to a drawing

Dist (di) – Measures the distance between two points in coordinate units

Distantlight – Creates a distant light

Divide (div) – Divides and entity into equal parts

Donut (do) – Creates a donut

Dsviewer – Brings the **Aerial View** window on screen

Dtext (dt) – Creates dynamic text. Text appears in drawing area as it is entered

Dxbin – Brings the **Select DXB File** dialog on screen

Dxfin – Brings the **Select File** dialog on screen

Dxfout – Brings the **Save Drawing As** dialog on screen

Ellipse (el) – Creates an ellipse

Erase (e) – Erases selected entities from a drawing

Exit – Ends a drawing session and closes AutoCAD 2009

Explode (x) – Explodes a block or group into its various entities

Explorer – Brings the **Windows Explorer** on screen

Export (exp) – Brings the **Export Data** dialog on screen

Extend (ex) – To extend an entity to another

Fillet (f) – Creates a fillet between two entities

Filter – Brings the **Object Selection Filters** dialog on screen

Gradient – Brings the **Hatch and Gradient** dialog on screen

Group (g) – Brings the **Object Grouping** dialog on screen

Hatch – Allows hatching by the *entry* responses to prompts

Hatchedit (he) – Allows editing of associative hatching

Help – Brings the **AutoCAD 2009 Help: User Documentation** dialog on screen

Hide (hi) – To hide hidden lines in 3D models

Id – Identifies a point on screen in coordinate units

Imageadjust (iad) – Allows adjustment of images

Iimageattach (iat) – Brings the **Select Image File** dialog on screen

Imageclip – Allows clipping of images

Import – Brings the **Import File** dialog on screen

Insert (i) – Brings the **Inert** dialog on screen

Iinsertobj – Brings the **Insert Object** dialog on screen

Isoplane (Ctrl/E) – Sets the isoplane when constructing an isometric drawing

Join (j) – Join lines which are in line with each other or arcs which are from the same centre point

Laycur – Changes layer of selected objects to current layer

Laydel – Deletes and purges a layer with its contents

Layer (la) – Brings the **Layer Properties Manager** dialog on screen

Layout – Allows editing of layouts

Lengthen (len) – Lengthens an entity on screen

Limits – Sets the drawing limits in coordinate units

Line (l) – Creates a line

Linetype (lt) – Brings the **Linetype Manager** dialog on screen

List (li) – Lists in a text window details of any entity or group of entities selected

Load – Brings the **Select Shape File** dialog on screen

Ltscale (lts) – Allows the linetype scale to be adjusted

Measure (me) – Allows measured intervals to be placed along entities

Menu – Brings the **Select Customization File** dialog on screen

Menuload – Brings the **Load/Unload Customizations** dialog on screen

Mirror (mi) – Creates an identical mirror image to selected entities

Mledit – Brings the **Multiline Edit Tools** dialog on screen

Mline (ml) – Creates mlines

Mlstyle – Brings the **Multiline Styles** dialog on screen

Move (m) – Allows selected entities to be moved

Mslide – Brings the **Create Slide File** dialog on screen

Mspace (ms) – When in PSpace changes to MSpace

Mtext (mt or t) – Brings the **Multiline Text Editor** on screen

Mview (mv) – To make settings of viewports in Paper Space

Mvsetup – Allows drawing specifications to be set up

New (Ctrl+N) – Brings the **Select template** dialog on screen

Notepad – For editing files from the Windows **Notepad**

Offset (o) – Offsets selected entity by a stated distance

Oops – Cancels the effect of using **Erase**

Open – Brings the **Select File** dialog on screen

Options – Brings the **Options** dialog to screen

Ortho – Allows ortho to be set ON/OFF

Osnap (os) – Brings the **Drafting Settings** dialog to screen

Pagesetup – Brings either the **Page Setup Manager** on screen

Pan (p) – Drags a drawing in any direction

Pbrush – Brings **Windows Paint** on screen

Pedit (pe) – Allows editing of polylines. One of the options is **Multiple** allowing continuous editing of polylines without closing the command

Pline (pl) – Creates a polyline

Plot (Ctrl+P) – Brings the **Plot** dialog to screen

Point (po) – Allows a point to be placed on screen

Polygon (pol) – Creates a polygon

Polyline (pl) – Creates a polyline

Preferences (pr) – Brings the **Options** dialog on screen

Preview (pre) – Brings the print/plot preview box on screen

Properties – Brings the **Properties** palette on screen

Psfill – Allows polylines to be filled with patterns

Psout – Brings the **Create Postscript File** dialog on screen

Purge (pu) – Purges unwanted data from a drawing before saving to file

Qsave – Saves the drawing file to its current name in AutoCAD 2009

Quickcalc (qc) – Brings the **QUICKCALC** palette to screen

Quit – Ends a drawing session and closes down AutoCAD 2009

Ray – A construction line from a point

Recover – Brings the **Select File** dialog on screen to allow recovery of selected drawings as necessary

Recoverall – Repairs damaged drawing

Rectang (rec) – Creates a pline rectangle

Redefine – If an AutoCAD command name has been turned off by **Undefine**, **Redefine** turns the command name back on

Redo – Cancels the last **Undo**

Redraw (r) – Redraws the contents of the AutoCAD 2009 drawing area

Redrawall (ra) – Redraws the whole of a drawing

Regen (re) – Regenerates the contents of the AutoCAD 2009 drawing area

Regenall (rea) – Regenerates the whole of a drawing

Region (reg) – Creates a region from an area within a boundary

Rename (ren) – Brings the **Rename** dialog on screen

Revcloud – Forms a cloud-like outline around objects in a drawing to which attention needs to be drawn

Ribbon – Brings the ribbon on screen

Ribbonclose – Closes the ribbon

Save (Ctrl+S) – Brings the **Save Drawing As** dialog box on screen

Saveas – Brings the **Save Drawing As** dialog box on screen

Saveimg – Brings the **Render Output File** dialog on screen

Scale (sc) – Allows selected entities to be scaled in size – smaller or larger

Script (scr) – Brings the **Select Script File** dialog on screen

Setvar (set) – Can be used to bring a list of the settings of set variables into an AutoCAD Text window

Shape – Inserts an already loaded shape into a drawing

Shell – Allows MS-DOS commands to be entered

Sketch – Allows freehand sketching

Solid (so) – Creates a filled outline in triangular parts

Spell (sp) – Brings the **Check Spelling** dialog on screen

Spline (spl) – Creates a spline curve through selected points

Splinedit (spe) – Allows the editing of a spline curve

Status – Shows the status (particularly memory use) in a Text window

Stretch (s) – Allows selected entities to be stretched

Style (st) – Brings the **Text Styles** dialog on screen

Tablet (ta) – Allows a tablet to be used with a pointing device

Tbconfig – Brings the **Customize** dialog on screen to allow configuration of a toolbar

Text – Allows text from the Command line to be entered into a drawing

Thickness (th) – Sets the thickness for the Elevation command

Tilemode – Allows settings to enable Paper Space

Tolerance – Brings the **Geometric Tolerance** dialog on screen

Toolbar (to) – Brings the **Customize User Interface** dialog on screen

Trim (tr) – Allows entities to be trimmed up to other entities

Type – Types the contents of a named file to screen

UCS – Allows selection of **UCS** (User Coordinate System) facilities

Undefine – Suppresses an AutoCAD command name

Undo (u) (Ctrl+Z) – Undoes the last action of a tool

View – Brings the **View** dialog on screen

Vplayer – Controls the visibility of layers in Paperspace

Vports – Brings the **Viewports** dialog on screen

Vslide – Brings the **Select Slide File** dialog on screen

Wblock (w) – Brings the **Create Drawing File** dialog on screen

Wmfin – Brings the **Import WMF** dialog on screen

Wipeout – Forms a polygonal outline within which all crossed parts of objects are erased

Wmfopts – Brings the **WMF in Options** dialog on screen

Wmfout – Brings the **Create WMF File** dialog on screen

Xattach (xa) – Brings the **Select Reference File** dialog on screen

Xline – Creates a construction line

Xref (xr) – Brings the **Xref Manager** dialog on screen

Zoom (z) – Brings the zoom tool into action

3D commands

3darray – Creates an array of 3D models in 3D space

3dface (3f) – Creates a 3- or 4-sided 3D mesh behind which other features can be hidden

3dmesh – Creates a 3D mesh in 3D space

3dcorbit – Allows methods of manipulating 3D models on screen

3ddistance – Allows the controlling of the distance of 3D models from the operator

3dfly – Allows walkthroughs in any 3D plane

3dforbit – Controls the viewing of 3D models without constraint

3dmove – Shows a 3D move icon

3dorbit (3do) – Allows a continuous movement and other methods of manipulation of 3D models on screen

3dorbitctr – Allows further and a variety of other methods of manipulation of 3D models on screen

3dpan – Allows the panning of 3D models vertically and horizontally on screen

3drotate – Displays a 3D rotate icon

3dsin – Brings the **3D Studio File Import** dialog on screen

3dsout – Brings the **3D Studio Output File** dialog on screen

3ddwf – Brings up the **Export 3D DWF** dialog

3dwalk – Starts walk mode in 3D

anipath – Opens the **Motion Path Animation** dialog

Align – Allows selected entities to be aligned to selected points in 3D space

Ameconvert – Converts AME solid models (from Release 12) into AutoCAD 2000 solid models

Box – Creates a 3D solid box

Cone – Creates a 3D model of a cone

convertoldlights – Converts lighting from previous releases to AutoCAD 2009 lighting

convertoldmaterials – Converts materials from previous releases to AutoCAD 2009 materials

convtosolid – Converts plines and circles with thickness to 3D solids

convtosurface – Converts objects to surfaces

Cylinder – Creates a 3D cylinder

Dducs (uc) – Brings the **UCS** dialog on screen

Edgesurf – Creates a 3D mesh surface from four adjoining edges

Extrude (ext) – Extrudes a closed polyline

Flatshot – Brings the **Flatshot** dialog to screen

Freepoint – Point light created without settings

Freespot – Spot light created without settings

Helix – Construct a helix

Interfere – Creates an interference solid from selection of several solids

Intersect (in) – Creates an intersection solid from a group of solids

Light – Enables different forms of lighting to be placed in a scene

Lightlist – Opens the **Lights in Model** palette

Loft – Activates the **Loft** command

Materials – Opens the **Materials** palette

Matlib – Outdated instruction

Mirror3d – Mirrors 3D models in 3D space in selected directions

Mview (mv) – When in PSpace brings in MSpace objects

Pface – Allows the construction of a 3D mesh through a number of selected vertices

Plan – Allows a drawing in 3D space to be seen in plan (UCS World)

Planesurf – Creates a planar surface

Pointlight – Allows a point light to be created

Pspace (ps) – Changes MSpace to PSpace

Pyramid – Creates a pyramid

Renderpresets – Opens the **Render Presets Manager** dialog

Renderwin – Opens the **Render** window

Revolve (rev) – Forms a solid of revolution from outlines

Revsurf – Creates a solid of revolution from a pline

Rmat – Brings the **Materials** palette on screen

Rpref (rpr) – Opens the **Advanced Render Settings** palette

Section (sec) – Creates a section plane in a 3D model

Shade (sha) – Shades a selected 3D model

Slice (sl) – Allows a 3D model to be cut into several parts

Solprof – Creates a profile from a 3D solid model drawing

Sphere – Creates a 3D solid model sphere

Spotlight – Creates a spotlight

Stlout – Saves a 3D model drawing in ASCII or binary format

Sunproperties – Opens the **Sun Properties** palette

Torus (tor) – Allows a 3D torus to be created

Ucs – Allows settings of the UCS plane

–render – Can be used to make rendering settings from the command line. Note the hyphen (−) must precede **render**

Sweep – Creates a 3D model from a 2D outline along a path

Tabsurf – Creates a 3D solid from an outline and a direction vector

Ucs – Allows settings of the UCS plane

Union (uni) – Unites 3D solids into a single solid

View – Creates view settings for 3D models

Visualstyles – Opens the **Visual Styles Manager** palette

Vpoint – Allows viewing positions to be set from x,y,z entries

Vports – Brings the **Viewports** dialog on screen

Wedge (we) – Creates a 3D solid in the shape of a wedge

Xedges – Creates a 3D wireframe for a 3D solid

Internet tools

Etransmit – Brings the **Create Transmittal** dialog to screen

Publish – Brings the **Publish** dialog to screen

APPENDIX A

Some set variables

Introduction

AutoCAD 2011 is controlled by a large number of set variables (over 770 in number), the settings of many of which are determined when making entries in dialogs. Some are automatically set with *clicks* on tool icons. Others have to be set at the command line. Some are read-only variables which depend upon the configuration of AutoCAD 2011 when it originally loaded into a computer (default values). Only a limited number of the variables are shown here.

A list of those set variables follows which are of interest in that they often require setting by *entering* figures or letters at the command line. To set a variable, enter its name at the command line and respond to the prompts which arise.

To see all set variables, *enter* **set** (or **setvar**) at the command line:

```
Command:enter set right-click
SETVAR Enter variable name or ?: enter ?
Enter variable name to list <*>: right-click
```

And an **AutoCAD Text Window** opens showing a list of the first of the set variables. To continue with the list press the **Return** key when prompted and at each press of the **Return** key, another window opens.

To see the settings needed for a set variable *enter* the name of the variable at the command line, followed by pressing the **F1** key which brings up a **Help** screen, *click* the search tab, followed by *entering* set variables in the **Ask** field. From the list then displayed the various settings of all set variables can be read.

Some of the set variables

ANGDIR – Sets angle direction. **0** counterclockwise; **1** clockwise
APERTURE – Sets size of pick box in pixels
AUTODWFPUBLISH – Sets **Autopublish** on or off
BLIPMODE – Set to **1** marker blips show; set to **0** no blips
COMMANDLINE – Opens the command line palette
COMMANDLINEHIDE – Closes the command line palette
COPYMODE – Sets whether **Copy** repeats

> **Note**
>
> **DIM** variables – There are over 70 variables for setting dimensioning, but most are in any case set in the **Dimension Styles** dialog or as dimensioning proceeds. However one series of the **Dim** variables may be of interest

DMBLOCK – Sets a name for the block drawn for an operator's own arrowheads. These are drawn in unit sizes and saved as required

DIMBLK1 – Operator's arrowhead for first end of line

DIMBLK2 – Operator's arrowhead for other end of line

DRAGMODE – Set to **0** no dragging; set to **1** dragging on; set to **2** automatic dragging

DRAG1 – Sets regeneration drag sampling. Initial value is 10

DRAG2 – Sets fast dragging regeneration rate. Initial value is 25

FILEDIA – Set to **0** disables **Open** and **Save As** dialogs; set to **1** enables these dialogs

FILLMODE – Set to **0** hatched areas are filled with hatching; set to **0** hatched areas are not filled; and set to **0** and plines are not filled

GRIPS – Set to **1** and grips show; set to **0** and grips do not show

LIGHTINGUNITS – Set to **1** (international) or **2** (USA) for photometric lighting to function

MBUTTONPAN – Set to **0** no *right-click* menu with the Intellimouse; set to **1** Intellimouse *right-click* menu on

MIRRTEXT – Set to **0** text direction is retained; set to **1** text is mirrored

NAVVCUBE – Sets the **ViewCube** on/off

NAVVCUBELOCATION – Controls the position of the **ViewCube** between top right (**0**) and bottom left (**3**)

NAVVCUBEOPACITY – Controls the opacity of the **ViewCube** from **0** (hidden) to **100** (dark)

APPENDIX B

NAVVCUBESIZE – Controls the size of the **ViewCube** between **0** (small) and **2** (large)

PELLIPSE – Set to **0** creates true ellipses; set to **1** polyline ellipses

PERSPECTIVE – Set to **0** places the drawing area into parallel projection; set to **1** places the drawing area into perspective projection

PICKBOX – Sets selection pick box height in pixels

PICKDRAG – Set to **0** selection windows picked by two corners; set to **1** selection windows are dragged from corner to corner

RASTERPREVIEW – Set to **0** raster preview images not created with drawing; set to **1** preview image created

SHORTCUTMENU – For controlling how *right-click* menus show:

0 all disabled; **1** default menus only; **2** edit mode menus; **4** command mode menus; **8** command mode menus when options are currently available. Adding the figures enables more than one option

SURFTAB1 – Sets mesh density in the M direction for surfaces generated by the **Surfaces** tools

SURFTAB2 – Sets mesh density in the N direction for surfaces generated by the **Surfaces** tools

TEXTFILL – Set to **0** True Type text shows as outlines only; set to **1** True Type text is filled

TiILEMODE – Set to **0** Paperspace enabled; set to **1** tiled viewports in Modelspace

TOOLTIPS – Set to **0** no tool tips; set to **1** tool tips enabled

TPSTATE – Set to **0** and the Tool Palettes window is inactive; set to **1** and the Tool Palettes window is active

TRIMMODE – Set to **0** edges not trimmed when **Chamfer** and **Fillet** are used; set to **1** edges are trimmed

UCSFOLLOW – Set to **0** new UCS settings do not take effect; set to **1** UCS settings follow requested settings

UCSICON – Set **OFF** UCS icon does not show; set to **ON** it shows

Ribbon panel tool icons

Introduction

The ribbon panels shown are those which include tools described in the chapters of this book. Panels and tools which have not been used in the construction of illustrations in the book have not been included. If a tool in a panel has not been described or used in this book, the icon remains unnamed in the illustrations below. Where flyouts from a panel include tools icons, the flyouts have been included with the panels. Flyouts appear when an arrow to the right of the panel name is *clicked*. Where the names of tool icons have been included in the panels, the names have not been added to the illustrations as labels being deemed unnecessary.

2D Drafting and Annotation ribbon

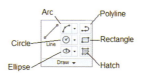

Fig. A3.1 The **Home/Draw** panel

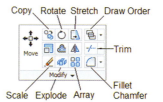

Fig. A3.2 The **Home/Modify** panel

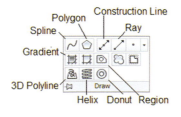

Fig. A3.3 The **Home/Draw** panel flyout

Fig A3.4 The **Home/Modify** panel flyout

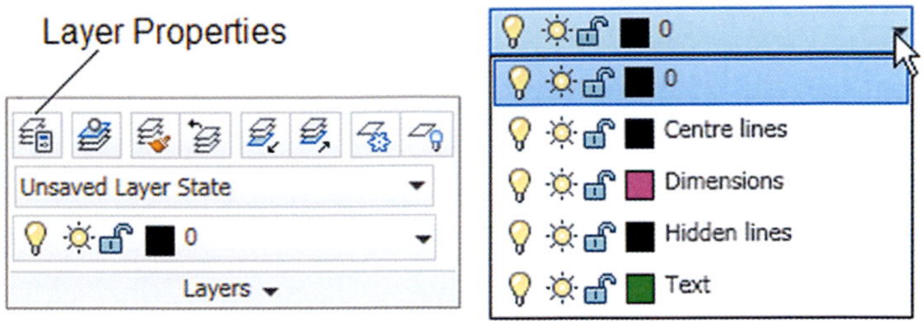

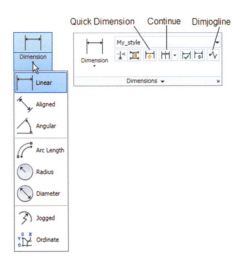

Fig. A3.5 The **Home/Layers** panel with the Layers drop-down menu

Fig. A3.6 The **Annotate/Dimensions** panel

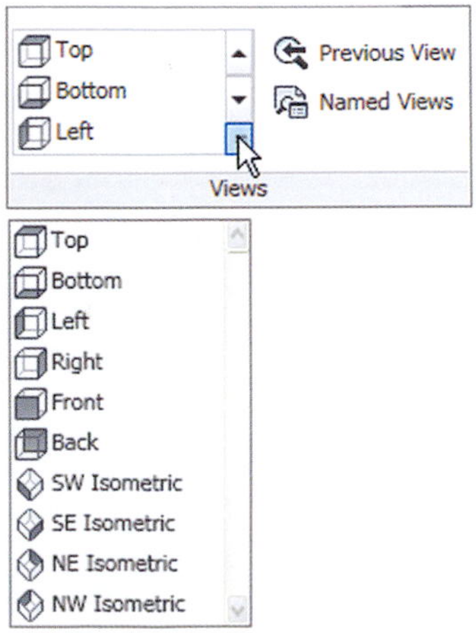

Fig. A3.7 The **View/Views** panel

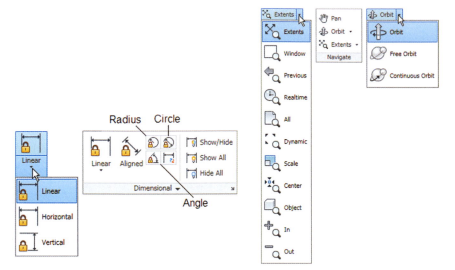

Fig. A3.8 The **Parametric/Dimensions** panel

Fig. A3.9 The **View/Navigate** panel

Fig. A3.10 The **View/Palettes** panel

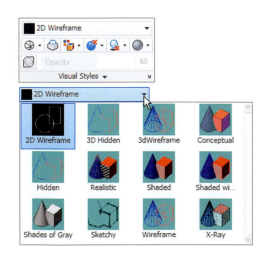

Fig. A3.11 The **View/Visual Styles** panel

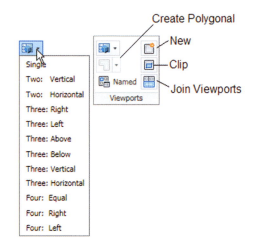

Fig. A3.12 The **View/Viewports** panel

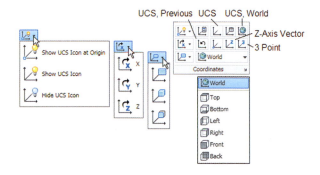

Fig. A3.13 The **View/Coordinates** panel

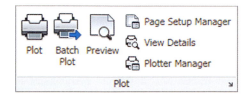

Fig. A3.14 The **Output/Plot** panel

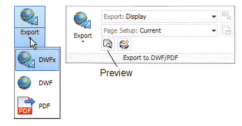

Fig. A3.15 The **Output/Export to DWF/PDF** panel

3D Modeling ribbon

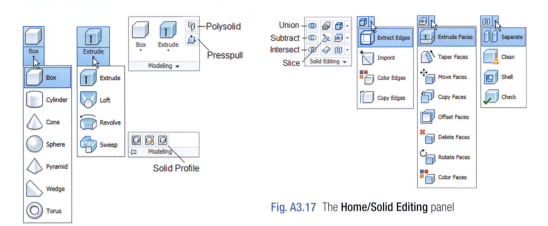

Fig. A3.17 The **Home/Solid Editing** panel

Fig. A3.16 The **Home/Modeling** panel and its flyout

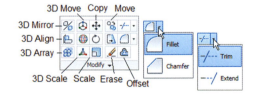

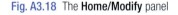

Fig. A3.18 The **Home/Modify** panel

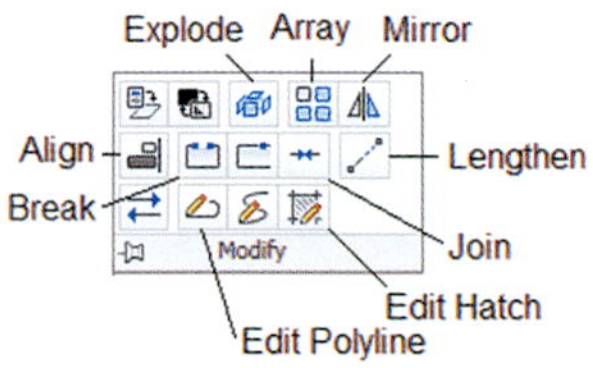

Fig. A3.19 The **Home/Modify** flyout

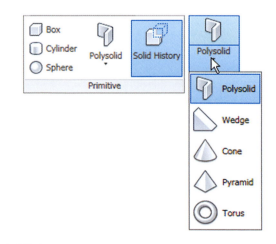

Fig. A3.20 The **Solid/Primitive** panel

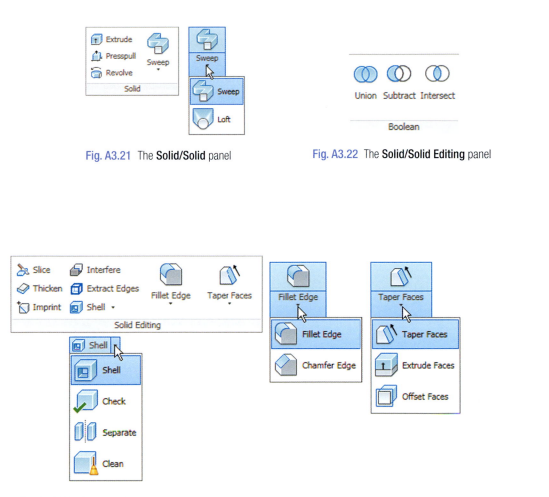

Fig. A3.21 The **Solid/Solid** panel

Fig. A3.22 The **Solid/Solid Editing** panel

Fig. A3.23 The **Solid Boolean** panel

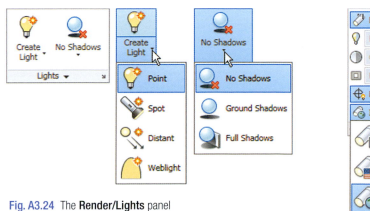

Fig. A3.24 The **Render/Lights** panel

Fig. A3.25 The **Render/Lights** flyout

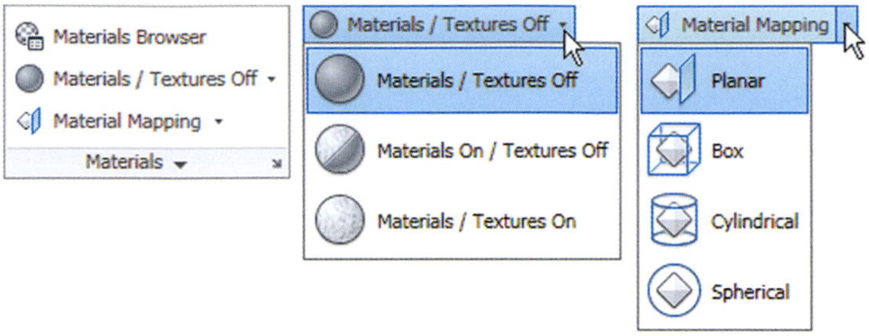

Fig. A3.26 The **Render/Materials** panel

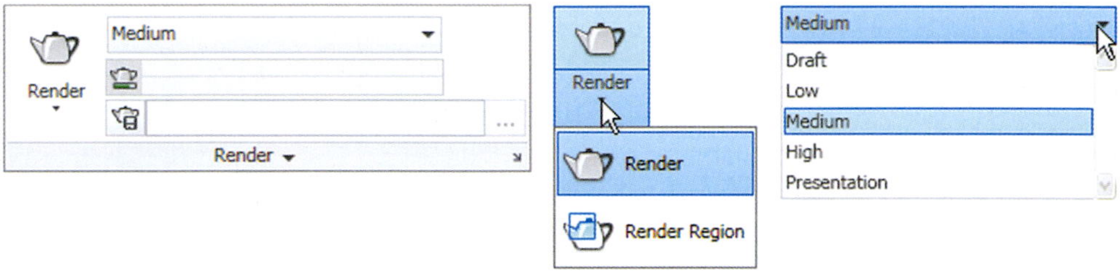

Fig. A3.27 The **Render/Render** panel

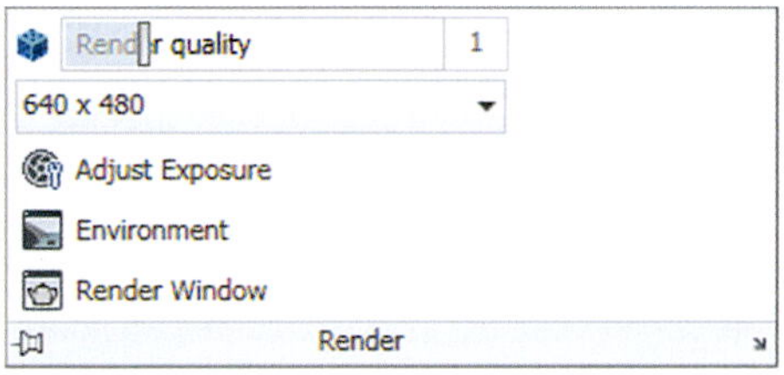

Fig. A3.28 The **Render/Render** flyout

Index